Effects of Welding on Health IV

An up-dated (December 1980-June 1982) literature survey and evaluation of the data recorded since the publication of the first report, to understand and improve the occupational health of welding personnel.

By
Winifred G. Palmer

Submitted by:
Tracor Jitco
1776 E. Jefferson Street
Rockville, MD 20852

Submitted to:
SAFETY AND HEALTH COMMITTEE
AMERICAN WELDING SOCIETY
550 N.W. LeJeune Road
Miami, FL 33126

International Standard Book Number: 0-87171-230-X

American Welding Society, 550 LeJeune Road, Miami, FL 33126

This report is published as a service and convenience to the welding industry and is the product of an independent contractor (Tracor Jitco) which is solely responsible for its contents. The materials in this report have not been independently reviewed or verified and are only offered as information. AWS assumes no responsibility for any claims that may arise from the use of this information. Users must make independent investigations to determine the applicability of this information for their purposes.

Printed in the United States of America

Contents

Personnel

The author of the report by Tracor Jitco was:

Winifred G. Palmer

AWS Research Committee

A.N. Ward, Chairman	Caterpillar Tractor Company
K.L. Brown, Vice Chairman	Lincoln Electric Company
M.E. Kennebeck, Jr., Secretary	American Welding Society
J.S. Gorski	Kemper Insurance Companies
E. Mastromatteo	INCO Limited

Acknowledgements

The American Welding Society gratefully acknowledges the time and effort expended by the members of the Research Committee and the financial support of the program by industry contributions.

Supporting Organizations

Air Products and Chemicals, Inc.
Airco Welding Products
Allis-Chalmers
Alloy Rods Division, The Chemetron Corporation
AWS Detroit Section
AWS New Orleans Section
Arcos Corporation
The Binkley Company
Caterpillar Tractor Company
Chicago Bridge and Iron Company
Grove Manufacturing Company, Division of Kidde, Inc.
General Electric Company
The Heil Company
Hobart Brothers Company
Huntington Alloys, Inc.
Lincoln Electric Company
Miller Electric Manufacturing Company
National-Standard Company
A.O. Smith Corporation
Teledyne-McKay, Inc.
Trinity Industries, Inc.
Truck Trailer Manufacturers Association
Walker Stainless Equipment Company
Weld Tooling Corporation

Many other organizations have made contributions to support the ongoing program from May 1979 to the present.

Preface

This literature review has been prepared for the Safety and Health Committee of the American Welding Society to provide an assessment of current knowledge of the effects of welding on health, as well as to aid in the formulation of research projects in this area, as part of an ongoing program sponsored by the Society. Previous work has included studies of the fumes, gases, radiation, and noise generated during various forms of arc welding (see Bibliography). Conclusions based on this review and reccomendations for further research are presented in the introductory portions of the report. Section 1 summarizes recent studies of the occupational exposures. Section 2 contains information related to the human health effects of exposure to byproducts of welding operations. Section 3 discusses studies of the effects of welding emissions on laboratory animals and *in vitro* cell systems. Referenced materials are available from Tracor Jitco, Inc.

Introduction

The health of workers in the welding environment is a major concern of the American Welding Society. To stay abreast of this subject, the health literature is periodically reviewed and published in the report *Effects of Welding on Health.* Three volumes have been published to date (Refs. 135, 143, and 144); the first covered data published prior to 1978, while the latter two covered the periods 1978 to May, 1979 and June, 1979 to December 1980, respectively. The current report included information that was published between December, 1980 and June, 1982, and should be read in conjunction with the previous volumes for a comprehensive treatment of the literature on the *Effects of Welding on Health.*

Included in this volume are studies of the characteristics of welding emissions that may have impact on the control technologies necessary to protect the welding worker (Section 1). Considerable discussion is devoted to ozone which may become a greater problem to welders as improved ventilation and decreased fume exposures reduce the rate of degeneration of this hazardous gas. Much recent research has focused on chromium and nickel, since exposure to certain chemical forms of these metals may cause serious chronic health problems.

In keeping with previous volumes, the health studies are organized according to the organ system affected. The respiratory tract, the primary route of entry of welding fumes and gases into the body, also is a major target organ of a number of components of these emissions. Acute (e.g., metal fume fever, cadmium poisoning) as well as potential chronic respiratory effects (e.g., emphysema, cancer) of welding emissions are of concern. The latter are far less well understood and whether or not there is an excess risk of cancer from these exposures has not been established. Continued research in the form of epidemiologic studies, investigations with laboratory animals, and *in vitro* genotoxicity studies will help to resolve this question.

Executive Summary

A problem inherent in research concerning the health effects of welding is that there is a great deal of variability in both welding processes and in working conditions which makes it difficult to perform studies on homogeneous populations of sufficient size to permit statistical analysis. These variations may be less critical when examining causes of acute physiological responses to welding exposures (e.g., metal fume fever, burns, photokeratitis). However, the association between chronic exposures to welding emissions and disease conditions whose causes are less well understood, or which occur at a low incidence, remains ambiguous in many cases.

The Respiratory Tract

Of the five metals (Cr, Ni, As, Be, and Cd) that have been shown by epidemiologic studies to be related to an increased cancer incidence in workers in certain industries, chromium and nickel are present in significant quantities in fumes released during welding certain metals (e.g., stainless steel and nickel alloys). However, only some chemical compounds containing these metals are carcinogenic, and it is not known if these compounds are present in welding fumes. The available epidemiologic evidence is insufficient to determine whether or not welders are at an increased risk of developing lung cancer from exposure to chromium and nickel in welding fumes.

Most of the reports of the effects of welding on health which appeared during the period of this review dealt with effects of welding on the respiratory tract. Metal deposits have frequently been found in welders' lungs by chest X-rays and at autopsy. The use of magnetopneumography enables rough estimates to be made of the quantities of metals retained in the lungs. However, the association between this retention of metals and impaired lung function is not clear. For this reason, it is important to develop more accurate methods for the determination of lung function. Of the six studies using pulmonary function tests of welders which appeared during this report period, three indicated impaired lung function among welders, and three reported no differences between welders and controls.

Four epidemiologic studies of the lung cancer incidence among welders have appeared in the recent literature. Two of these indicated an elevated lung cancer risk. One of these two studies focused on welders of stainless steel, but the size of the study cohort was too small for the results to be considered conclusive. The second was a death record study of welders selected from the rosters of a trade union in Seattle, Washington. A significant increase (SMR*=174) in the lung cancer rate was found when deaths that occurred more than 20 years from the initial date of exposure were considered. A study of welders exposed to nickel-rich welding fumes did not indicate an elevated lung cancer risk; however, the investigator emphasized the need for a follow-up study of this cohort to allow more time for the appearance of tumors. There is a great need for further well-planned epidemiologic studies of welders. Because of the high cost of these studies and the limited resources available to the welding community, Stern suggested

*Standardized mortality ratio

that such studies focus on exposure to welding processes in which it is suspected that genotoxic or carcinogenic materials, such as nickel and chromium compounds, are released (Ref. 122).

Other Organ Systems

Reports of recent health studies that examined the effects of welding on organ systems other than the respiratory tract are highlighted below.

-- With the exception of cutaneous burns and erythema of the neck which were more common in welders, no significant differences between the frequency of skin and eye disorders were found between welders and nonwelders from the same fabrication facility. Good safety and health measures were practiced in this plant.

-- An excess of inflammation of the oral mucous membranes, including periodontal disease, was observed in two separate studies of welders.

-- No differences were observed in the morphology, number, and motility of sperm produced by welders and by workers in other occupational groups.

Biological Monitoring of Exposure to Welding Emissions

The development of methods for the accurate determination of personal exposures are needed to establish and measure the effectiveness of control technologies. In conjunction with devices that collect personal air samples for determining the welder's actual exposure, biological monitoring affords a means for estimating the actual dose of contaminants taken up by the body. Such tests are useful for the support of population and epidemiologic studies and for the evaluation of the effectiveness of industrial hygiene measures introduced to reduce worker exposure. Finally, biological monitoring during routine medical examinations may play an important part in alerting the occupational physician to individuals who may be exposed to unacceptable levels of welding emissions.

The importance of this area of research is witnessed by the number of studies that have been published on this subject. Studies using magnetopneumography have demonstrated that the amount of materials retained in the lung correlates with the number of years spent welding, the extent of siderosis, and the relative quantity of fumes generated by the welding process. In other work, the relationship between urine concentrations and breathing zone levels was linear for fluorine, suggesting that this element may be useful for monitoring the exposure of welders who use basic-coated electrodes. Although urine chromium levels tended to increase during the work week, this metal nevertheless may be useful for estimating exposures to fumes generated by welding of stainless steel. Urine nickel levels tend to fluctuate widely and are apparently not a useful measure of nickel exposures.

Technical Summary

The Exposure

Fumes

The major objectives of research on fume emissions are to evaluate health hazards, to develop welding methods which produce less toxic fumes, and to enable estimation of air exchange rates required to bring fume concentrations to acceptable levels. Total fume emissions are greatest with shielded metal arc welding and flux cored arc welding and vary with the electrode used and welding parameters.

To determine the airflow requirements of ventilation systems, Alekseeva *et al.* examined the emissions released during gas tungsten arc welding of manganese-copper alloys (Ref. 4), copper alloys containing nickel or zinc (Ref. 3), and eight different welding wires containing from 2 to 70% nickel (Ref. 5). Pokhodnya *et al.* (Ref. 102) found that the rate of manganese vaporization from rutile or basic electrodes increased with the alkalinity of the slag. Oleinchenko *et al.* (Ref. 98) determined that fluoride emissions from electrode coatings increased with the moisture of the coating during automatic aluminum welding.

A large portion of the solid phase of welding aerosols consists of respirable particles. Particles may exist as single entities or as chains or agglomerates (Refs. 12, 23, 51, 70 and 81). When present, chromates appear to be condensed on the surface of metal oxide particles (Refs. 70 and 81).

Two major interlaboratory studies are currently examining methods for the determination of soluble and insoluble hexavalent chromium in welding emissions (Refs. 11 and 51). Variations in the methods of sample collection, storage, and extraction procedures may affect the analytical results (Refs. 11, 49, and 128).

Gases

The sources of toxic gases, e.g., nitrogen oxides, ozone, and carbon monoxide, during welding are discussed. Johansson (Ref. 71) reported that the addition of helium or hydrogen to shield gases reduced the ozone levels released during GMAW or GTAW of stainless steel, and Smars (Ref. 117) reported that the addition of nitric oxide, but not helium, significantly reduced ozone levels during GTAW of aluminum. However, Farwer (Ref. 39) argued that because of its toxicity and stability, the NO_2 generated from nitric oxide in shield gases may be a greater hazard than the ozone whose formation it prevents. The presence of magnesium, but not silicon, in aluminum alloys significantly reduces the production of ozone during GMAW of aluminum (Ref. 117).

Production Coatings

Primers or paints on metal surfaces may contribute significant quantities of formaldehyde, carbon monoxide, hydrogen cyanide, and other organic vapors to welding emissions (Ref. 21). These vary primarily with the characteristics of the binders used in the coatings.

Electromagnetic Radiation

Since ultraviolet radiation in the vicinity of 270 to 280 nm is most likely to be absorbed by and damage the genetic material (DNA), Bartley *et al.* (Ref. 14) investigated the intensity of radiation with these wavelengths generated by GTAW and GMAW of aluminum. They found that, unlike its effect on ozone levels, magnesium in aluminum alloys effects an increase in ultraviolet radiation between 270 and 280 nm.

Pabley and Keeney (Ref. 99) showed that reflective gold polycarbonate filter plates in welders helmets more effectively reduced exposure to infrared radiation than did green glass filter plates. Sliney *et al.* (Ref. 116) developed design criteria for a semitransparent curtain which would enable safe bystander viewing of the welding area and increase visibility within the welding booth.

Noise

Hermanns (Ref. 65) compared the noise levels produced by different welding processes and found that, while most produced at least occasional noise above 85 dB, plasma cutting was the noisiest.

Effects of Welding on Human Health

Respiratory System

Using magnetopneumography, the quantity of particles retained in the lungs of welders in the asbestos (Ref. 31) and shipyard (Ref. 42) industries was found to be significantly higher than that in workers not employed in welding. Of workers using a variety of common welding methods, stainless steel welders employed in SMAW had the highest quantities and those using GTAW had the lowest quantities of retained metals in the lung (Refs. 73 and 80). Welders performing SMAW of mild steel had intermediate levels of retained metals.

Two cases of pulmonary fibrosis were reported in aluminum arc welders (Refs. 62 and 134). Cigarette smoke and other exposures may also have contributed to the disease. Gobatto *et al.* (Ref. 45) examined lung tissue from 17 autopsied welders. Fibrosis was present in more than half, and siderosis was present in all of the lungs examined. Areas of inflammation or fibrosis consistently had associated iron deposits which led the investigators to speculate that the metal or associated toxins, or both, were involved in the disease process. Zober (Ref. 147) reviewed and analyzed reports of 47 cases of pulmonary fibrosis in welders published between 1955 and 1979. He concluded that, at most, only 20 cases could have been related to the welding experience.

Hayden *et al.* (Ref. 60) found that welders in the engineering industry had slightly more absences due to upper respiratory tract infections than did controls, but absenteeism resulting from other diseases did not differ from control groups. McMillan and Molyneux (Refs. 83 and 84) found no differences in the absentee patterns from respiratory tract diseases in nonsmoking welders and controls. However, they did note that smoking welders tended to have slightly more absences from lower respiratory tract disease than did the nonwelding controls who smoked tobacco products.

The results of lung function tests of welders are inconsistent. No significant differences between the results of pulmonary function tests of welders and controls were found by Schneider and Rebohle among a population of shipyard workers (Ref. 111) or by Hayden *et al.* (Ref. 60) among workers in the engineering industry. Zober examined 40 welders and 40 age matched controls and, with the exception of siderosis detected by chest X-rays, no abnormalities in pulmonary function were identified that could be related to welding (Ref. 145). However, Akbarkhanzadeh (Ref. 2) found that welders in the shipyard industry had impaired lung function, and Cavatorta (Ref. 25) reported changes in respiratory function which were related to the duration of the welding experience. Ohmori *et al.* (Ref. 97) found a slight decrease in peak flow rate and maximum expiratory flow rate at 75% vital capacity which was related to ozone exposures during GMAW. According to Stern (Ref. 122), some of these inconsistencies may result from population dynamics, whereby workers with pulmonary difficulties may change occupations, and to bystander effects whereby some welders may be exposed to toxic substances from nonwelding sources at the place of employment.

Lung Cancer

Polednak (Ref. 103) compared the causes of mortality among welders of nickel alloys with a separate cohort of workers who welded mild steel, stainless steel, and aluminum. The number of deaths from all causes as well as diseases of the respiratory tract, including cancer, were the same in both groups and did not differ significantly from the general U.S. population. Because the period of time following the initial welding exposures may have been too short to rule out an increased respiratory cancer risk in welders exposed to nickel-containing fume, Polednak (Ref. 103) and Gibson (Ref. 44) stressed that follow-up studies should be performed with this important cohort as the number of years from the initial exposure to welding fumes increases.

Beaumont and Weiss (Ref. 16) examined the number of deaths from lung cancer among 3247 welders selected from the rosters of a trade union in Seattle, Washington. A significant increase (SMR = 174) in the lung cancer rate was found when deaths that occurred more than 20 years from the initial date of exposure were considered. Sjogren (Ref. 115) noted an increased risk of lung cancer (3 deaths observed versus 0.68 expected) among a population of 234 stainless steel welders in Northern Sweden. The results of a death certificate study by Ahonen (Ref. 1) indicated that the risk of lung cancer was the same for welders and for workers in other occupations.

Ear and Hearing

A survey of the effectiveness of earmuffs and earplugs indicated that welders and platers who use earplugs may suffer less hearing damage than those using earmuffs (Ref. 38). The investigators speculated that their results may reflect the greater ease with which earmuffs can be removed and replaced by the wearer.

Eye and Vision

A Bureau of Labor Statistics Survey showed that welding and cutting injuries represented 0.5% of the Workmen's Compensation cases in the United States in 1978. Of these injuries 67% involved the eye, 1/3 of which were welder's flash. Emmett *et al.* (Ref. 36) found no difference between visual acuity or frequency of ocular abnormalities between welders, nonwelders exposed to welding fumes, and workers who were not exposed to welding fumes. All workers were employed for a comparable time period at the same facility.

A case of a pterygium (ultraviolet-induced growth on the outer surface of the eye) was reported in a 56 year-old former welder (Ref. 68). It is unlikely that this resulted from work-related exposure to ultraviolet radiation (Refs. 96 and 126).

Skin

Scars from skin burns were found on 45%, and ultraviolet-induced dermatitis in 8% of welders during a survey of Soviet shipyard workers (Ref. 131). The results of a survey performed in a Danish hospital indicated that welders suffered from burns more frequently than workers in other occupations (Ref. 112). A study in a U.S. facility revealed more cutaneous scars resulting from thermal and mechanical injuries in welders and machinists than in other workers (Ref. 36).

In the latter study, facial erythema, actinic elastosis, and premalignant and malignant skin lesions were found with the same frequencies in welders and nonwelders. Erythema of the neck was more frequent in welders. The investigators noted that this was a relatively young population and the absence of excess premalignant and malignant skin lesions may not be conclusive due to the relatively short period of time following initial welding exposures.

Cardiovascular System

An experimental apparatus that simulated the welding experience was designed and used in a laboratory setting to determine changes in heart rate while carrying out typical welding procedures (Ref. 130). The variation in heart rate was greatest when the "welder" altered his posture to change electrodes. This variation was greater with a stooped (downhand welding) position than with a standing (overhead welding) or a crouched position.

Nervous System

Chandra *et al.* found that urine manganese and serum calcium levels were elevated in welders who worked with manganese-containing electrodes and had positive neurological symptoms (tremors and deep brisk tendon reflexes in the extremities) (Ref. 27). The investigators suggested that serum calcium levels may be of value in diagnosing early stages of manganese poisoning.

Musculoskeletal System

An incidence of about 15 to 20% was found for supraspinatus tendinitis in shipyard workers (Ref. 63). The number of years of welding experience and the rated level of shoulder muscle load could not be correlated with the development of supraspinatus tendinitis, but these factors could not be excluded by the results of this study. Kadefors *et al.* (Ref. 72) measured the myoelectric activity in individual muscles while eight basic welding postures were assumed by human subjects in an experimental setting. All positions caused localized muscle fatigue, but the overhead welding position was potentially the most damaging to shoulder muscles. This study indicated that the assumption of certain positions may reduce the total muscle load in different welding situations. Tumakov and Grigorev (Ref. 132) designed an exercise program aimed at reducing tension in muscles subjected to static stress to be followed by welders suffering from lumbosacral radiculitis.

Urogenital Tract

No differences were observed in the morphology, number, and motility of sperm produced by welders and workers in other occupational groups (Ref. 76). This and an earlier study (Ref. 58), indicate that welding has no effect on fertility.

A study by Alsbirk *et al.* (Ref. 6) detected no changes in kidney function in welders of stainless steel.

Teeth and Oral Cavity

An excess of inflammation of the oral mucous membranes, including periodontal disease, was observed by Wulf and Seefeld (Ref. 140) and Melekhin and Agarkov (Ref 89). To investigate the cause of complaints by underwater electric arc welders of a metallic taste in their mouth and damage to their dental fillings, Rockert, Christensson, and Orthendal (Ref. 28 and 108) studied the underwater effects of electric currents on dental amalgams. They found that the current through the amalgam was dependent on its surface area and distance from the electrode.

Metal Fume Fever and Allergic Reactions

Nine cases of metal fume fever in welders were reported in Mexico (Ref. 18). Five of the affected welders had pulmonary abnormalities detected by X-rays.

Zugarman (Ref. 148) described a case of a welder with an unusually severe allergic response to chromium. An *in vitro* study of the response of isolated white blood cells to manganese and fluorine indicated that 50% of the welders tested were sensitive to either manganese or fluorine and 25% were sensitive to both elements. None of the nonwelding controls was responsive (Ref. 89). Since the physiological response was not examined by standard allergy tests, the importance of the *in vitro* results cannot be evaluated.

Biochemical Changes

Elevated blood lead levels were observed in welders of steel coated with anti-corrosive paints (Ref. 46). Elevated blood lead levels, decreased blood hemoglobin levels, and increased urine levels of delta-aminolevulinic acid were found in 22 tank workers in Egypt who were exposed to lead fumes for over 22 years, but had no overt clinical signs of lead poisoning (Ref. 90). These biochemical alterations are typical for lead poisoning. Evidence of abnormal liver function was also indirectly indicated by the increases in the serum enzymes SGOT, SGPT, and LDH.

Elevated plasma nickel levels were found in shipyard workers (Ref. 47). Urine nickel levels, which fluctuated throughout the day, were less accurate markers for nickel exposure than plasma nickel levels. Sjogran (Ref. 114) examined the relationship between breathing zone levels and urine concentrations of fluorine, chromium, and nickel. The relationship between urine and breathing zone levels was linear for fluorine. Urine chromium content increased throughout the work week even though airborne levels were relatively constant. The relationship between airborne and urinary nickel levels was nonlinear.

Human Fatality

A human fatality was reported to be due to inhalation of cadmium fumes from the use of an oxyacetylene torch with silver solder (Ref. 82). The solder, which contained 20% cadmium, was not labeled as such by the manufacturer.

Toxicological Investigations in Animals and Cell Cultures

Animal Studies

The extent of the inflammatory response to foreign particles in the lung can be monitored by measuring the number of migratory leukocytes (free cells), the levels of surfactant, or the hydrolytic enzyme activity in the respiratory tract. Using these parameters, White *et al.* (Ref. 137) found that particles generated by SMAW of stainless steel using a rutile electrode elicited a slightly greater inflammatory pulmonary reaction than did particles from SMAW of mild steel using basic or rutile electrodes. Speculating that the response to the stainless steel particles may have been largely due to the chromates, they performed a follow-up study in which the inflammatory response elicited by the water-soluble and insoluble fractions of particles from the stainless steel welding system was compared with that elicited by potassium dichromate. Although the chromate salt enhanced all of the parameters studied, the effects observed are general responses to pulmonary irritants, and it cannot be concluded from these studies that chromate is responsible for the inflammatory response elicited in the lung by fumes from welding stainless steel.

Repeated administration of fumes from coated electrodes led to the appearance of fibrotic materials around aggregates of welding particles in the lungs of guinea pigs and rats (Ref. 7). The fibrotic response was also evidenced by the elevation of hydroxyproline in the lungs. Insufficient data were presented for a determination of the factors responsible for the fibrotic response.

The importance of examining the toxic effects of complete mixtures rather than isolated components of industrial pollutants was demonstrated by Il'nitskaya and Kalina (Ref. 69). These investigators found that repeated inhalation by rats of a mixture of aluminum oxide and welding gases (1 mg/m^3 ozone and 3.3 mg/m^3 NO_x) caused less pulmonary damage than did inhalation of the welding gases alone. The authors suggested that the toxicity of the gases was reduced by adsorption to the aluminum oxide particles.

English *et al.* (Ref. 37) compared the distribution of NiOand $NiCl_2$ throughout the body following administration of the salts by intratracheal instillation to rats. The water-insoluble NiO salt was distributed more slowly from the lung to other organs than was the water-soluble $NiCl_2$ but was eventually found in the same organs as was $NiCl^2$. The investigators concluded that NiO gradually becomes converted in the lung and the adjacent lymph nodes to a more water-soluble form which allows the nickel to become distributed throughout the body.

Kawata *et al.* (Ref. 77) examined the acute effects on laboratory animals of inhalation of high concentrations of fumes from basic-coated electrodes. Fumes from electrodes containing lithium were less toxic than fumes from comparable lithium-free basic electrodes.

In Vitro Studies

Using the sister chromatid exchange assay, Niehbuhr *et al.* found that nickel-rich welding fumes cause DNA alterations (Refs. 94 and 95). The genotoxicity of the fumes was comparable to that of nickel sulfate. The authors concluded that the water-solubility of nickel compounds in welding fumes is an inadequate indicator of its potential toxicity.

White *et al.* (Ref. 138) compared the cytotoxicity of fumes generated by SMAW of mild steel using either basic or rutile-coated electrodes and by SMAW of stainless steel using a rutile-coated electrode. Stainless steel fumes inhibited the growth of a rapidly proliferating human cancer cell line. Fumes from basic electrodes caused the most hemolysis of red blood cells, and all three welding methods caused about a 30% decrease in the viability of cultured macrophages.

Conclusions

A number of unanswered questions persist concerning the effects of welding on health. Most of these problems have been reviewed in past volumes of this series. (*Effects of Welding on Health,* Volumes I, II, and III) and many are also problems encountered in other industries where dusts and fumes are prevalent. Some of the major health-related questions and recommendations for addressing these questions are presented.

Pulmonary Function Tests

Pulmonary function tests may be inadequate for detecting lesions in the smaller airways. The reliability of these tests for general medical screening of welders and prospective welders requires further study.

Cancer Incidence

No general agreement is apparent in the findings of the many epidemiologic studies as to whether or not there is an elevated cancer incidence among welders. This may be, in part, because the work conditions are exceedingly variable from workplace to workplace, and studies of welders working under different conditions may not be comparable. Another difficulty is that, unless the cancer risk is markedly elevated, a very large exposed population is needed to unequivocally determine a disease incidence. A clear definition of the work conditions, exposure levels, welding materials, and general health (including medical and smoking histories) may allow the integration of results from different investigations, and thereby increase the size of the population studied. Coordination of studies sponsored by different associations (as suggested by Stern, Ref. 123) and adoption of standard protocol designs (Refs. 122 and 146) would help to achieve this goal.

Chromium and Nickel Carcinogenicity

Of the solid components released during welding of clean metals, only nickel and hexavalent chromium have been implicated as human carcinogens. It is not known whether the forms of nickel and chromium present in welding fumes are carcinogenic. Only one epidemiologic study has indicated an elevated respiratory cancer risk in welders of stainless steel, and in this work, the size of the study cohort was quite small (Ref. 115). As Stern suggested (Ref. 122), epidemiologic studies should focus on welding shops in which "hot spots" for cancer may exist. In other words, those welding conditions (e.g., welding of stainless steel or high nickel alloys) which are suspected of releasing genotoxic or carcinogenic materials should be high priority targets for investigation.

Mutagenicity Tests

It is a widely held belief that at least some occupational and environmental exposures which cause cancer do so by causing damage to DNA. As discussed earlier, many *in vitro* tests assess the ability of test substances to damage DNA. A major problem with *in vitro* tests is that they ignore the immune defense mechanisms present in live animals as well as the ability of metabolic systems to detoxify foreign chemicals. Furthermore, when complex emissions are being examined by *in vitro* tests, they do not reflect the fact that, in animal systems, there may be a selective uptake (especially when airborne mixtures enter the body through the respiratory tract), absorption, and organ distribution of the various components of the mixture. What is apparently toxic in an *in vitro* system, therefore, may be relatively innocuous to the whole animal.

A technique, which has been used with workers in other industries (Refs. 10 and 141), that partially overcomes these difficulties involves examining cultured white blood cells collected from exposed workers and appropriate controls for evidence of any DNA damage incurred *in vivo* following natural exposures. This technique, combined with the analysis of personal air samples, discriminates between components that may or may not enter or remain in the body and allows a direct examination of effects on exposed workers. A disadvantage of this method is that it only examines one organ system and may not pick up the chromosomal damage produced by toxins that only affect specific target organs. Nonetheless, because it more accurately reflects the actual workplace exposures, the examination of the effects of welding exposures on the genetic material of white blood cells collected from welders is a promising method for dealing with a difficult problem.

Interactions Between Components of Welding Fumes, Gases, and Vapors

With few exceptions, threshold limit values and other exposure levels are based on the known toxicologic effects of single elements or compounds. Thus, synergistic toxicologic interactions between chemicals are not taken into account and for certain combinations of exposures or components within mixtures, the combined effects may be greater than the additive effects of the individual components. In this case, permissible levels may not be sufficient to protect the worker. On the other hand, various components of exposures may act to reduce the toxic effects of other components of the exposure, and an unnecessary financial burden may be placed on industry in conforming to overly stringent regulations.

Although *in vitro* assays may be useful for examining certain types of multiple effects, animal tests are probably most useful for detecting synergists or antagonists because these reactants may exert their effects through entirely different organ systems than the toxin whose action they are affecting (e.g., one component of a mixture might suppress the immune system which could, by reducing the body's defense mechanisms, increase the effectiveness of a carcinogen in a different organ such as the liver). Little is known about interactions between components of industrial emissions, and this is a most important area for future research.

Another type of interaction between components of mixtures is that suggested in the study by Il'nitskaya and Kalina (Ref. 69). In that case, pulmonary lesions resulting from exposure to welding gases (ozone and NO_x) were greater in the absence of alumina than in its presence. This was thought to be due to the adsorption of the gases onto the alumina particles. Health professionals should consider the effect that the design of new control technologies to eliminate or reduce some components of an exposure could have on the remainder of the components. Another example of this is that reduction of fumes during welding may lead to an increase in ozone levels. Thus, during the introduction of any new technologies, or changes in existing technologies, the effects on the whole system must be considered.

Ergonomics

Welding places a strain on the pectoral girdle, back, and shoulder muscles. Although, as discussed earlier, the use of specific body positions while welding may decrease some of the static load on affected muscles (Ref. 72), these strains cannot be eliminated. As emphasized by Spelbrink (Ref. 119), there is a real need for the development of mechanical aids which can perform much of the static holding functions in welding and thereby reduce the muscle strain experienced in this type of work.

Health and Safety

Once identified, factors responsible for adverse health effects are likely to be controllable. In only one of the reports of workplace studies cited above (Ref. 36), it was stated that data were collected in a work environment in which there was a strong emphasis on worker safety and health. In this case, skin and eye lesions among welders were minimal. As discussed by Hinrichs (Ref. 66), studies of the health effects of welding in workplaces where recommended safe practices are used would be most important in determining the effectiveness of these practices.

Effects of Welding on Health IV

1. The Exposure

Welders may be exposed to the fumes and gases of welding emissions, radiation, noise, and organic vapors released during the welding of primed or degreased metals. The emissions produced by welding of clean, umprimed metals consist of fumes containing metal oxide particles, and gases which are primarily oxides of nitrogen (NO_x), ozone (O_3) and carbon monoxide. The concentrations of these components vary greatly with the welding method. Although all may be present in differing amounts during welding by the various methods, in general, the greatest quantities of fumes are produced during shielded metal arc welding (SMAW) and flux cored arc welding (FCAW); ozone is produced in the highest concentrations during gas metal arc welding (GMAW), especially of aluminum; and the NO_x are most prevalent during gas burning welding processes. Carbon monoxide levels may be significantly elevated during welding of metal coated with primers which contain organic binders and during gas metal arc welding with a CO_2 shield.

1.1 Fumes

The fumes, which consist primarily of minute metal oxide particles, constitute the solid phase of welding emissions and originate mainly from the welding consumables. The quantity and constituents of fumes depend on the welding process, the electrode type, the welding current and voltage, and other parameters. The objectives of much of the research performed on welding fume generation are to develop low fume electrodes and also to enable estimation of air exchange rates required to bring the fume concentrations below established permissible levels.

Fume generation is generally measured either on the basis of the total quantity of fume emission per unit time, the fume generation rate or FGR (gram/min), or the relative fume formation rate, RFFR, which is the quantity of emissions produced when a certain mass of consumable electrode (or filler metal) is melted (weight fume/weight electrode) (Ref. 8). Similar measures are used in the Scandinavian countries where the fume generation rate is symbolized as E (mg/sec) and the relative fume formation rate is symbolized as R and is measured in terms of mg fume/kg electrode.

1.1.1 Effects of Voltage and Current

Of the commonly used welding methods, with the exception of GMAW of aluminum, SMAW generates the highest fume levels and gas tungsten arc welding (GTAW) the lowest (Refs. 70 and 133). For SMAW, total fume emission, FGR, increases with welding current. Since the increased fume emissions reflect an increased melting rate, the welding current has little influence on the ratio of fume emissions to electrode consumption (RFFR). However, both FGR and RFFR increase with increasing voltage. Johansson *et al.* explained that this may be due to the longer arc which, by evoking greater air turbulence, facilitates the penetration of oxygen through the protective gas shield with resulting increases in metal oxide formation (Ref. 70).

Eichhorn *et al.* (Ref. 35) reported that raising the welding current has a greater effect on fume emission from acid or basic-coated electrodes than on that from acid rutile-coated electrodes. Other factors, such as polarity, may alter fume generation rates. The extent of the alteration, and the polarity which produces the highest fumes, are dependent on the electrode type. Alternating current tends to produce less fumes than direct current (Ref. 35).

In GMAW with an argon shield, RFFR is independent of voltage, once the voltage is high enough to produce a stable metal spray. However, using a mixture of argon and CO_2, RFFR increases with voltage once a stable spray arc is established. With a CO_2 shield, the total fume emission, FGR, increases greatly as the voltage is increased (Ref. 70).

Castagna and Spagnoli (Ref. 24) reported that small differences in the arc voltage and current resulted in major differences in the concentration of fumes and ozone in the breathing zones of welders using argon shielded GMAW with continuous feed electrodes composed of copper-coated mild steel. By increasing the power from 280A and 28V to 300A and 30V, the concentration of ferric oxide was reduced five-fold, the oxides of manganese, zinc, copper, and nitrogen were reduced approximately ten-fold, and the ozone level increased by a factor of eight.

1.1.2 Effects of Electrode Composition

Several investigators compared the quantity of fumes generated with different types of electrodes of electrode-base metal combinations (Refs. 3, 4, 5, 35, 88 and 102). Eichhorn (Ref. 35) reported that acid rutile-coated electrodes produce less fumes than do acid or basic-coated electrodes. Of the electrodes tested, the cellulose-coated electrodes used in open air pipeline construction generate the highest quantities of fumes. In general, the total quantity of fumes (FGR) generated increases with the diameter of the electrode. However, in relation to the total amount of metal deposited, fume emission decreases with increasing electrode diameter.

To determine the airflow requirements of ventilation systems, Alekseeva *et al.* examined the emissions released during GTAW of manganese-copper alloys with an argon shield (Ref. 4) and of copper with a nitrogen shield (Ref. 3). In the first study, the quantity of emissions during welding of three different base metals composed of alloys containing copper, small amounts of aluminum, nickel, and iron and approximately 43, 53, and 63% manganese, respectively, were studied. The two alloys containing the higher percentage of manganese also contained small amounts of zinc. The filler wire was composed of a copper alloy containing 43% manganese. The manganese content of the fumes was directly related to the manganese content of the base metal and it varied from 2.4 to 26.2 mg/M^3 in front of the shield and 0.8 to 2.2 mg/M^3 behind the shield. Manganese, copper and zinc oxide, but no aluminum or nickel oxides, were present in the weld emissions. For every alloy examined, manganese oxide, but not copper or zinc oxide, reached levels behind the face shields that were higher than the permissible exposure levels in the USSR. The alloys containing zinc gave off more total solid materials than did that alloy which lacked zinc. In the former case, zinc represented close to 50% of the total particulate material in the welding emissions.

In the second study (Ref. 3), emissions from filler and base metal combinations with varying quantities of nickel and zinc were compared. As in the previous system, the total emissions were greatest when zinc was present in the copper alloy. In systems with little or no zinc, copper oxide was the primary component of the emissions that had to be considered during the design of ventilation systems. When zinc alloys were used, zinc oxide levels were sufficiently high that they had to be considered as well.

A third study by Alekseeva *et al.* (Ref. 5) examined the concentrations of nickel in front of the welder's shield during GTAW with an argon shield of eight samples of welding wire which contained 2.2 to 70% nickel. The welding conditions were kept constant during this test. Nickel was not found in the welding fumes (lower limit of detection - 0.2 μg/ml sample) when the nickel content of the welding wire was less than 5%. When the welding materials contained nickel in concentrations equal to or greater than 5%, the amount of nickel in the fumes increased with the nickel content of the welding wire.

Pokhodnya *et al.* (Ref. 102) compared the release of iron and manganese from slags from rutile or basic electrodes of differing alkalinities. The rate of manganese vaporization, but not that of iron, increased with the alkalinity of the slag.

Oleinchenko *et al.* (Ref. 98) found that the quantities of soluble (NaF and KF) and insoluble (AlF_3) fluoride compounds, as well as the quantity of gaseous fluoride (HF), increased with the moisture content of the electrode coating (44% cryolite, 38% KCl, 15% NaCl, 3% silicon) during automatic aluminum welding. Precalcination of the welding electrodes at 380 to 420°C significantly reduced emissions of the fluoride compounds.

Buki and Feldman (Ref. 22) developed equations for estimating the concentrations of metal compounds in the emissions generated by gas-shielded welding of steels and copper and aluminum alloys. Tests were performed which demonstrated that the calculated concentrations of a number of metal oxides (nickel, aluminum, manganese, copper, silicon, chromium, and iron) generated during welding of steel with strip electrodes and during argon-shielded arc welding and plasma arc welding were in the same range as those determined experimentally. The authors concluded that their equations can be put to practical use, for example, to obtain the initial data for calculating the requirements of ventilation systems.

1.1.3 Particles

The size and shape of airborne particles govern their aerodynamic behavior within the respiratory tract which, in turn, determines whether or not they are deposited in the lung. Spherical particles greater than 5 μm in aerodynamic diameter will generally be trapped within the upper respiratory tract (e.g. nasal passages, sinuses, trachea) and expelled, whereas particles between 0.1 and 5 μm in diameter are considered to be respirable; that is, they can be inhaled and retained within the deep lung. Particles smaller than 0.1 μm may not be slowed down sufficiently by mechanisms such as impaction and may remain within the airstream to be removed during the expiration cycle. The aerdynamic behavior of fibers such as asbestos differ from that of spherical particles, and fibers up to 20 μm in length have been found within the lung.

In general, particles present in welding fumes are spherical. Irregular shaped particles may also be present in the air in welding shops, but these usually arise from grinding and other mechanical operations (Ref. 13). Most reports indicate that the greatest number of particles in welding fumes tend to have diameters which range from less than 0.1 μm up to 0.5 μm (Refs. 23, 35, 51, 70 and 88). Johansson *et al.* (Ref. 70) reported that the mass median diameters (the total mass of the particles with diameters larger than the mass median diameter is equal to the total mass of the particles with diameters smaller than the mass median diameter) of particles produced by different welding processes ranged from less than 0.25 μm for fumes produced by GTAW to 0.3 to 0.6 μm for GMAW and SMAW of stainless steel.

The particles may exist in welding fumes as unique, single entities or they may be present as agglomerates or chains (Refs. 12, 23, 51, 70 and 88). Carsey (Ref. 23) reported that particles in the solid phase of fumes from SMAW of mild steel consist of two fractions; the first is represented by single, nonagglomerated particles 0.05 to 0.1 μm in diameter, and in the second, the particles are agglomerated into aggregates 0.25 to 0.5 μm in diameter. Particles from SMAW of stainless steel tended to be less heavily agglomerated and had a more even size distribution, which was in the range of 0.05 to 0.4 μm.

Other investigators (Refs. 51 and 70) reported that many of the airborne particles from SMAW of stainless steel are present as chains or agglomerates of smaller particles. The count median diameter (half the total number of particles are larger in diameter and half smaller than the count median diameter) was reported to be 0.15 μm by Gray *et al.* (Ref. 51) and to be less than 0.1 μm by Johansson *et al.* (Ref. 70). Gray *et al.* noted the presence of some large, "glossy coated spherical particles" in the SMAW fumes that were up to 10 μm in diameter. They also found that particles from GMAW tend to be spherical and are present in fumes both as independent entities as well as in chains. They postulated that the chains and free particles form by different mechanisms. Differences in the observations of these investigators are probably due to variations in the welding processes rather than the analytical techniques employed.

A number of investigators have applied relatively new or experimental techniques to examine the elemental composition of particles in welding fumes. These techniques included X-ray fluorescence, in which particle are irradiated with X-rays and the resultant fluorescing X-ray spectra are analyzed (Ref. 23); Particle-Induced X-ray Emission (PIXE) in which the X-ray spectra emitted by particle bombarded with protons are analyzed (Refs. 70 and 86); and scanning electron microscopy coupled with Energy Dispersive X-ray Analysis (EDXA) in which X-ray spectra from particles irradiated with an electron beam are examined (Refs. 81 and 133). In one study, Electron Spectrocopy for Chemical Analysis (ESCA), which also provides information on the oxidation state of elements, was used (Ref. 70). With this method, only the elements in a very thin layer on the particle surface can be analyzed.

Malmqvist *et al.* (Ref. 86) and Johansson *et al.* (Ref. 70) used Particle-Induced X-ray Emission (PIXE) to analyze the elements present in particles of welding aerosols. Thirteen different combinations of electrodes and base metals with SMAW, GMAW, or GTAW were used. With low alloy steels, fluorine, potassium, calcium (these three were with SMAW only), iron and manganese were most abundant. With stainless steel electrodes, chromium, nickel, and low levels of molybdenum were found. Aluminum oxide was the main component of aluminum welding aerosols and only very low quantities of a few other elements were detected. The fumes generated by SMAW, GMAW, and GTAW were separated by a cascade impactor and the elements in each particle size range were analyzed. No marked differences were observed in the distribution of elements.

With SMAW using stainless steel electrodes, chromium, manganese and iron increased and potassium and fluoride decreased with increasing voltage. The difference may arise from the fact that potassium and fluorine originate from the electrode coating while the other elements generally originate from the melting metal core (Ref. 86). The elemental composition of fumes from GMAW of stainless steel did not vary with voltage (Refs. 35 and 70).

Koponen *et al.* (Ref. 81), using scanning electron microscopy with EDXA to examine particles from SMAW of stainless steel, found that nickel may exist as discrete particles. Chromates appeared to be present as condensates on the surface of metal oxide particles. Using ESCA, Johansson *et al.* (Ref. 70) found that 60 to 100% of the chromium on the particle surfaces is in the hexavalent state with SMAW of stainless steel, while less than 15% of that on the particles generated by GMAW is hexavalent. Washing the SMAW particles with a pH 7.4 buffer significantly reduced the chromium on the outer surface.

Barfoot *et al.* (Ref. 13) studied the variation with time of day of the chemical and physical characteristics of particulates and pollutants in a welding shop using a streak sampling method. Attempts were made to correlate analyses with activity in the shop. The analyses were done in a welding shop ventilated through ceiling ducts. Occasional melting and grinding operations were also performed on the premises as well as argon arc, oxyacetylene welding, and brazing of a variety of metals. Samples were collected at a distance of about 2 m from the welding table and 1 m above the ground. PIXE and scanning electron microscopy with EDXA were used for the determinations.

Amounts of elements such as aluminum, silicon, zinc, sodium, potassium, and calcium varied dramatically (up to three orders of magnitude) with time, and pollution levels were, on average, ten times higher during the day than at night. Most of the particles were less than 3 μm, and a large fraction were betwen 0.5 and 1.5 μm in diameter. Iron and aluminum were occasionally found in discrete spherical particles composed of a single element, and with a mean diameter of about 1 μm. Silicon and aluminum were sometimes found in needle-like shapes of up to 2 μm in length. The other particles contained combinations of two or more elements. Many irregular shaped particles were found, some of which could be associated with the grinding operations.

1.1.4 Chromium

Much emphasis has been placed on the development of reliable methods for determining the concentration of hexavalent chromium (Cr-VI) in welding fumes, primarily because of reports that it may be carcinogenic for humans (Ref. 91). Although an increased risk has only been demonstrated in workers in the chromium industry and not in welders, new stringent threshold limit values (TLV) have been proposed by NIOSH (Ref. 91) which would have industry-wide effects. Totally reliable methods for the extraction of Cr-VI from all types of welding fumes have not yet been established (Ref. 52). The potential need for compliance with strict standards and concern for worker safety make it necessary to be able to evaluate confidently the levels of hexavalent chromium in welding fumes.

An "interlaboratory round robin" study, organized by the American Welding Society and described by Andrews and Hanlon (Ref. 11), evaluated technique developed by Blakeley and Zatka (Ref. 19) for the determination of Cr-VI in welding fumes. Similar to the technique described by Thomsen and Stern (Ref. 129), the Blakeley-Zatka method relies on the extraction of chromates with sodium carbonate at high pH to avoid the reduction of Cr-VI to Cr-III by Fe-II, which can occur under the acid conditions described in the method recommended by NIOSH (Ref. 91).

In this multiple laboratory study, bulk fume samples collected from GMAW and SMAW methods with stainless steel consumables were analyzed using a slight modification of the Blakeley-Zatka method; the final chromium determination was made by atomic absorption spectrophotometry. The bulk of the chromium in GMAW fumes was not hexavalent; these fumes contained very low quantities of water soluble and insoluble Cr-VI. On the other hand, most of the chromium in SMAW fumes was soluble Cr-VI, with smaller quantities of non-Cr-VI and very low quantities of insoluble Cr-VI. A second round of testing was done with a fraction of SMAW fumes collected by polyvinyl chloride membrane filters. The average value for water-insoluble Cr-VI was slightly higher, and that of non-Cr-VI slightly lower than that obtained by the participating laboratories for the bulk sample. The results of the tests of SMAW fumes obtained by six of the laboratories are shown in Table 1.

Table 1
Chromium content of SMAW fumes collected on polyvinyl chloride membrane filters

	1	2	3	4	5	6	MEAN (Relative Standard Deviation)
Soluble Cr-VI	4.35	4.72	4.86	4.48	5.02	4.26	4.62 (6.5%)
Insoluble Cr-VI	0.81	0.12	0.36	0.37	0.19	0.34	0.365 (66%)
Non-Cr-VI	0.43	0.37	0.94	1.38	0.71	0.48	0.718 (54%)
Total Cr (Sum of above)	5.59	5.21	6.16	6.23	5.92	5.08	5.70 (8.6%)

Andrews and Hanlon Ref. 11

Andrews and Hanlon concluded that the potential for reduction of Cr-VI during sampling can be minimized by collecting samples on inert filters, storing samples under dessication, and performing analyses as soon as possible after collection. Alterations that can be produced in the Cr-VI content of fume samples by different collection and storage methods are further discussed by Gray *et al.* (Ref. 49) and by Thomsen and Stern (Ref. 128).

Preliminary results of a similar interlaboratory study, organized by the International Institute of Welding, were reported by Gray *et al.* (Ref. 51). Cr-VI was extracted from the fumes with a solution of Na_2CO_3 and NaOH. Some variation was observed between the results obtained by the four participating laboratories, as in the preceding study. Fumes from SMAW with low hydrogen or rutile covered electrodes had considerably more Cr-VI but less total chromium than did fumes from GMAW of stainless steel.

That there is substantially more Cr-VI in emissions from SMAW than from GMAW of stainless steel was also reported by other investigators (Refs. 114 and 133). In the latter case, the Cr-VI content of the fume may vary with the welding parameters; fumes produced by GMAW with short circuiting transfer arcs were reported to contain significantly more water soluble Cr-VI than fumes generated by spray arc conditions (Ref. 128). As described by Koponen *et al.* (Ref. 81), with SMAW of stainless steel using basic electrodes, the chromium in the slag becomes oxidized to Cr-VI with the oxidation of the slag forming elements and then reacts with alkali oxides from the electrode coating to form soluble chromates (e.g., $CaCrO_4$, K_2Cr_207) which condense on the surface of the slag and metal oxide particles. The absence of both oxygen and alkali oxides in welding with inert gas shields contribute to the absence of soluble Cr-VI in emissions generated by the GMAW method.

Because the carcinogenicity as well as other forms of biological activity may vary with the solubility of Cr-VI compounds in aqueous media (Ref. 91), the determination of water soluble Cr-VI is becoming routine in the analysis of stainless steel fume samples. Gray *et al.* (Ref. 52) discussed problems inherent in the available methods for the analysis of total chromium and the determination of soluble Cr-VI. These authors described a method which can more completely dissolve fume samples in preparation for analysis by atomic absorption spectrometry than can the standard methods which may leave a solid residue with some fume samples. They also demonstrated that, for the determination of the soluble Cr-VI compounds, different methods will lead to varying results, and they stressed the need for standardizing both the definition of Cr-VI solubility and methods for its determination.

1.2 Gases

The major gases that are generated during welding of umprimed metals include ozone, carbon monoxide, carbon dioxide, and the nitrogen oxides. These gases arise from interactions between ultraviolet radiation, intense heat or electric currents with normal atmospheric constituents, and also with some shielding gases (e.g., carbon dioxide). As described below, recent research has been performed on the generation of ozone, the nitrogen oxides, and carbon monoxide during welding.

1.2.1 Nitrogen Oxides

At temperatures greater than 1000° C, molecular nitrogen in the air is coverted into nitric oxide (NO). NO is readily oxidized in air at ambient temperatures to form nitrogen dioxide (NO_2). Although nitrogen oxides (NO_x) may be produced during most welding processes, the levels of NO_x are greatest during gas welding and plasma cutting (Ref. 107).

Press (Refs. 105 and 106) examined levels of nitrogen oxides produced by gas welding processes. As previously reported, (see *Effects of Welding on Health,* Volume III), his studies demonstrated that the quantity of nitrogen oxides produced is proportional to the length of the flame. A family of curves for different flame lengths was developed, showing the variation in the nitrogen oxides produced per unit time with different size nozzles. The quantity of nitrogen oxides increased with the size of the nozzle and the length of the flame. With small variations, the curves were similar for welding with oxyacetylene, propane, and natural gas. A practical comparison between the quantities of nitrogen oxides produced with different fuel gases could not be made because nozzles of the same size do not necessarily give comparable performances with different gases. In an experimental setting, where the same heat transfer to the workpiece and equal temperatures at the underside of the plate were achieved, the least amount of nitrogen oxides was produced with propane gas.

1.2.2 Ozone

Ozone is produced by the interaction of ultraviolet radiation of wavelengths lower than 200 nm with oxygen. Ozone is readily decomposed in air to molecular oxygen. The rate of decomposition is accelerated by metal oxide fume which also prevent its formation by blocking the transmission of ultraviolet light so that ozone presents little problem to welders using FCAW or SMAW.

Farwer (Ref. 39) reported that ozone does not present a hazard, even without local fume extraction, with GTAW of Cr-Ni steel or aluminum. With GTAW, a helium shield resulted in higher levels of ozone than did an argon shield. Significant quantities of ozone are generally formed during GMAW. The highest levels of ozone are produced by argon-shielded welding of aluminum (Ref. 133). With this welding process, ozone is more likely to be a problem, relative to permissible levels, than other emissions.

A factor which may reduce ozone levels is the presence of nitric oxide (NO) which reacts readily with ozone to produce molecular oxygen and nitrogen dioxide (NO_2).

$$NO + O_3 \rightarrow O_2 + NO_2$$

Because of this, the addition of small quantities of nitric oxide in shield gases has been recommended to reduce ozone levels during SMAW (Refs. 43 and 117). However, Farwer does not agree with this method and he argues that the reaction may not be rapid or complete enough to effectively reduce ozone levels. In support of this, he cites evidence that significant quantities of NO and O_3 may be found together in welding aerosols (Refs. 15 and 40). He suggests that another reason these two gases may coexist in the weld area is that NO_2 may be photochemically converted back to NO by the action of ultraviolet light. Since NO is readily converted to NO_2 which, because of its stability, represents a greater hazard than ozone, Farwer cautions strongly against the use of NO as an additive in shield gases to control ozone levels.

A difference between the generation of ozone and other welding emissions is that ozone forms not only in the immediate vicinity of the arc, but also further away from it, albeit in decreased quantities. According to Farwer, the addition of additives to the argon shield may actually change the spatial distribution of ozone production and although it may decrease the ozone in the vicinity of the arc, it may increase the ozone concentration in the breathing zone of the welder (Ref. 39).

In contrast to the report by Farwer, Johansson (Ref. 71) reported that the levels of ozone during GMAW and GTAW welding could be significantly reduced by inclusion of helium or hydrogen in the shield gases. For GTAW welding of stainless steel with an argon shield, the ozone level decreased as the concentration of hydrogen in the shield gas was increased from 0 to 5%. For GMAW welding of stainless steel, the ozone concentration was reduced by 15% when the shield gas was changed from 98% argon + 2% CO_2 to 63.8% argon + 3.2% CO_2 + 32% helium + 1% hydrogen. In this work (Ref. 71), measurements were made at various distances from the arc. The concentration of ozone in front of the welder's helmet, resulting from GMAW of aluminum, decreased markedly as the helium concentration of the argon shield was increased from 0 to 30%. Further reductions in ozone levels were far less dramatic as the helium concentration was increased to 100%. Differences between Farwer's data (Ref. 39) and that of Johansson (Ref. 71) cannot be resolved with the information available, but Johasson's work is in accord with other reports that shield gas additives can effectively reduce ozone levels (Refs. 41, 43 and 117).

Smars (Ref. 117) found that varying the amount of helium from 0 to 100% in the argon shield with GTAW of aluminum had only a minimal effect on the ozone produced. A slight increase in ozone levels in the region of the weld occured as helium increased from 0 to 30% of the shield gas, and a small decrease in the ozone level occurred as the helium concentration was increased from 70 to 100%. The nitrogen dioxide levels varied in the opposite direction. On the other hand, he reported that inclusion of nitric oxide in the shield gas significantly reduced ozone levels.

Smars (Ref. 117) also found that ultraviolet radiation and ozone, but not nitrogen dioxide increase with the arc length, and that as the arc current is increased from 150 to 300A, ozone levels reach a maximum between 225 and 250A. Welding rods of A1Si5 generated the most ozone followed closely by nonalloyed aluminum rods. The ozone levels generated by welding rods containing magnesium (AlMg3 and AlMg5) were approximately half that of rods containing silicon. Similarly, Frei *et al.* (Ref. 43) reported that ozone levels were much lower when gas metal arc welding base metals composed of AlMg5 than when welding AlSi5.

1.2.3 Carbon Monoxide

Carbon monoxide forms during the incomplete combustion of carbonaceous material and is generally not formed in significant quantities during SMAW of uncoated metals with the majority of electrodes. Carbon monoxide can be produced during GMAW with a CO_2 shield (Refs. 39 and 119), and occasionally elevated levels with peak readings of 150 ppm were recorded with this method (Ref. 133). Significant carbon monoxide levels can also form by combustion of the welding gases during oxyfuel gas welding.

1.3 Protective Coatings

Substances released during welding of primed or painted metals and from degreasing agents have, in recent years, been recognized as a potential source of toxic materials and have received increasing attention in health and research studies of the welding profession. The gases and organic vapors emitted from metals with protective coatings were examined by Bohme and Heuser (Ref. 21) using GTAW without a filler metal to minimize airborne contaminants from materials other than the coatings. Of the components measured in the welding emissions, carbon monoxide, the nitrogen oxides hydrogen cyanide (HCN), formaldehyde, and toluene diisocyanate (released from polyurethane binders) were found to be the most important when considered in relation to MAK (maximum safe workplace concentrations) values established in the Federal Republic of Germany (see Table 2 for TLV's and MAK values for some substances found in welding fumes). HCN and carbon monoxide, which arise from binders in the production coatings, were produced in the least amounts by ethyl silicate and the highest amounts by epoxy resins. The quantity of HCN, which was proportional to the amount of binder in the coating, was also low with alkyd resins and high with polyurethane resins. Although levels of nitrogen oxides were significant, they were not influenced by the coatings.

Levels of phenol released from phenol resins were low in relation to MAK values. Of the aldehydes detected, butyraldehyde and acetaldehyde were present in concentrations too low to be of concern in the workplace. On the other hand, formaldehyde, which has a lower MAK value (1 ml/m^3) than the other aldehydes, was present in significant amounts. Formaldehyde was not released in detectable quantities from binders containing zinc dust, which was attributed to the low percent of binders in zinc-containing coatings but, as the authors suggested, may also have been due to an inhibition of formaldehyde formation by the zinc.

The levels of gases were measured in front of and behind a face shield placed at a normal working distance from the arc. In front of the face shield, the MAK levels were exceeded for carbon monoxide, formaldehyde (this was high for only one of the coatings tested) and toluene disocyanate (this was three times the MAK level with the one polyurethane resin tested). Although none of the gases exceeded MAK levels behind the face shield, the investigators warned that care should still be exercised when welding over protective coatings since the thickness of the paint films often exceeds that suggested by the manufacturer, especially in corners and with manual painting. The resultant increase in emissions could easily exceed MAK values behind the face shield.

The fumes emitted from steel plates coated with 29 shop primers or components of primers (pigments - ferric oxide, zinc, aluminum, zinc chromate; extenders - barium sulfate, magnesium silicate, calcium carbonate; and six different binder systems) during GTAW welding were examined by Koerber and Fissan (Ref. 79). With the exception of primers containing zinc pigments, the total mass of the emissions was relatively low compared with those released from SMAW or GMAW with a CO_2 shield of unpainted steels. Zinc oxide represented the bulk of the emissions from zinc primers and, according to the investigators, the total quantity of emissions from these primers was similar to those typically released from SMAW or CO_2 welding of uncoated steels.

Low levels of lead and cadmium were released from all primers; the primers which produced the highest lead levels contained high contents of zinc or ferric oxide and the zinc-containing primers produced the highest cadmium levels. More of the particles generated from the primers had diameters greater than 0.3 μm than did those from uncoated steel. Most of the particles were smaller than 1 μm and their mass median diameter was approximately 0.1 μm.

During a field study of fumes released from SMAW and GMAW of non-alloy steel either unprimed or primed with a thin film of red shop primer containing the pigments zinc tetraoxychromate and iron oxide, Ulfvarson reported that iron, zinc, and chromium were released in greater amounts from primed steel, as would be expected (Ref. 133). However, manganese and copper also increased, even though these elements

Table 2
Guideline exposure levels in the United States (TVL) and maximum allowable concentration in the Republic of Germany (MAK) to substances that may be produced by welding

Substance [a]	TLV/TWA[b]	MAK Level [c,d]	Origin [d]
acetaldehyde	100 ppm	200 ml/m³	epoxy resin coating on the workpiece
acrolein	0.1 ppm	0.1 ml/m³	alkyd resin coating on the workpiece
antimony	0.5 mg/m³	0.5 mg/m³	alloying component
arsenic (soluble)	0.2 mg/m³	-	alloying component
beryllium	0.002 mg/m³	-	alloying component
cadimum	0.05 mg/m³	0.05 mg/m³	alloying component or anit-corrosion finish
carbon dioxide	5000 ppm	5000 ml/m³	GMAW with CO_2 shield
carbon monoxide	50 ppm	30 ml/m³	GMAW with CO_2 shield; production coatings, incomplete combustion
chromium-VI	0.05 mg/m³	-	principal alloying component in stainless steels
chromium (other)	0.5 mg/m³	-	
cobalt [e]	0.1 mg/m³	-	alloying component
copper (fume)	0.2 mg/m³	0.1 mg/m³	welding copper
fluorides	2.5 mg/m³	2.5 mg/m³	constituent of coatings, covering for rod electrode
formaldehyde [f]	2 ppm	1 ml/m³	epoxy resin coating on the workpiece
hydrocyanic acid	10 ppm	10 ml/m³	expoxy resin or polyurethane coating on the workpiece
iron oxide	5 mg/m³	8 mg/m³	welding of iron and steel
lead	0.15 mg/m³	0.1 mg/m³	alloying component or component of paint finishes
magnesium oxide	10 mg/m³	8 mg/m³	alloying component
manganese (dust and compounds)	5 mg/m³	5 mg/m³	constituent of many electrodes
(fumes)	1 mg/m³		
mercury (aryl & inorganic compounds)	0.1 mg/m³	0.01 ml/m³	anti-fouling paint used in shipbuilding
molybdenum (soluble)	5 mg/m³	5 mg/m³	alloying component
(insoluble)	10 mg/m³	-	
nickel (soluble)	0.1 mg/m³	-	alloying component, welding of nickel
nitric oxide	25 ppm	-	
nitrogen dioxide	3 ppm	5 ml/m³ (plus nitric oxide)	forms with virtually all welding processes but expecially with gas welding processes
ozone	0.1 ppm	0.1 ml/m³	GMAW, especially of aluminum
phenol	5 ppm	5 ml/m³	epoxy resin coating on the workpiece
phosgene	0.1 ppm	0.1 ml/m³	breakdown product of chlorinated hydrocarbons (degreasing agents)
phospine	0.3 ppm	0.1 ml/m³	anti-rust protection containing phosphate, impurity in acetylene
tin	2 mg/m³	2 mg/m³	welding of bronze
titanium (respirable)	5 mg/m³	8 mg/m³	alloying component and special purpose material
vanadium	0.05 mg/m³	0.1 mg/m³	alloying component
zinc (fume)	5 mg/m³	5 mg/m³	constituent of brass

a. Information in parentheses refer to Threshold Limit Values only.

b. Threshold limit values for chemical substances in work air adopted by the American Conference of Governmental Industrial Hygienists for 1982.

c. Maximum Safe Workplace Concentrations allowable in the Federal Republic of Germany.

d. Taken from Reidiger, Ref. 107.

e. The ACGHI has proposed a change in the TLV to 0.05 mg/m³ for dusts and fumes.

f. The ACGHI has proposed a change in the TLV to 1 ppm.

were not present in the primer. An explanation offered by the investigator for this observation was that larger diameter electrodes with a higher current are used for welding primed metals, releasing more metal oxides from the base metal. However, it was also noted that the thickness of the welded materials were not recorded, and a higher current would have been used had the thickness of the primed metal pieces been larger than that of the unprimed metals, which would have caused the same effect (Ref. 133).

Kireev *et al.* (Ref. 78) analyzed the emissions produced by GMAW with a CO_2 shield on steel plates coated with four different protective primers. Manganese oxides, styrene, and unsaturated hydrocarbons were the components of the emissions of greatest concern generated by welding of the Russian primer coats, EF-0121, VL-023, MS-067, and EP-057. (The composition of the primers was not stated.)

1.4 Electromagnetic Radiation

Electromagnetic radiation in the ultraviolet, visible, and infrared region of the spectrum is produced during most welding processes. Ultraviolet radiation shorter than 175 nm is readily absorbed by oxygen, and is therefore, dissipated close to the weld. Ultraviolet radiation of longer wavelengths is less readily dissipated and can be a problem with some welding processes. Ultraviolet radiation tends to be insignificant during SMAW because its transmission is readily blocked by the heavy fumes. Of the commonly used welding processes GMAW, especially of aluminum, produces the most intense ultraviolet radiation.

Bartley *et al.* (Ref. 14) measured the intensity of the ultraviolet radiation with wavelengths greater than 250 nm produced by helium shielded GTAW and argon-shielded GMAW of aluminum. In this work, they were most interested in radiation in the vicinity of 270 to 280 nm since ultraviolet radiation with these wavelengths is most likely to react with genetic material (DNA) and, by implication, may be the spectral region associated with skin cancer. Their results showed that with GMAW the use of consumable electrode wires with 5% magnesium produced markedly more radiation between 275 and 300 nm than did electrode wires without magnesium. With helium-shielded GTAW without filler metal, about ten times more ultraviolet radiation in this wavelength range was produced when welding aluminum plates with 2.5% magnesium than when magnesium was not present. The latter findings are in contrast with the work of other investigators (Refs. 39, 71 and 117) who found that ozone, which arises from the interaction of ultraviolet light with molecular oxygen, is less intense with GMAW of aluminum containing magnesium than when welding unalloyed aluminum. Since Bartley *et al.* (Ref. 14) measured only the intensity of ultraviolet radiation with wavelengths between 250 and 350 rm, the apparent conflict between this work and that of the other investigators cannot be resolved.

Exposure to the electromagnetic radiation produced by welding in the ultraviolet (greater than 175 rm), visible (especially in the blue-violet range), and infrared regions can cause severe skin and eye injuries. Glass or polycarbonate filter plates in welder's helmets can effectively prevent injury from most electromagnetic radiation (as well as metal spatter). However, according to Pabley and Keeney (Ref. 99), when materials (e.g., ferric oxide) which absorb infrared are incorporated into the filter plates, the absorbed infrared may be re-radiated from the plates into the area of the welder's eyes. Reflective metal coats (usually gold, silver, or aluminum) on the outer surface of the polycarbonate filter plate inhibit the absorption of infrared radiation and reflect much of it away from the plate. Pabley and Keeney compared the heat buildup on green glass filter plates and reflective gold coated polycarbonate plates during welding of safety plates; the helmet filter plate temperature was 216% and 54% higher than ambient with green glass and reflective polycarbonate filters, respectively. With the safety plates in place, the difference in the heat buildup on the back safety plate was far less dramatic, and there was little difference between the two types of filters. Thus, reflective polycarbonate filters may introduce a real advantage, in terms of infrared radiation exposure to the welder's eyes, only when the back safety plate is not in use.

To enable observation of welders for supervisory or safety purposes, to reduce the welder's sensation of confinement, and to increase visibility within the welding booth, Sliney *et al.* (Ref. 116) developed design criteria for a semitransparent curtain which would shield the bystander from hazardous radiation from the arc. The absorption of infrared radiation was not an important consideration since at the bystander's distance of 1 to 2 m from the arc, the intensity is too low to be damaging. Thus, the primary concern in the design of the curtain was to filter out ultraviolet radiation and the intense blue visible light from the arc that is not necessary for visibility. Incorporated into the curatin design are fluorescent dyes (to absorb ultraviolet radiation and reemit visible light) and submicron zinc oxide particles (to absorb short wave ultraviolet radiation, to serve as light scattering centers for the diffusion of transmitted light, to reduce the brightness of the arc spot, and to scatter the visible light returned to the curtain from the arc spot to increase general illumination of the work area.)

Unstable arc temperatures, variations in arc length, drafts, and shifts in air currents cause fluctuations in the electromagnetic radiation produced during welding. Gvozdenko *et al.* (Ref. 55) discussed the inadequacy of available instrumentation to measure accurately the total quantity of electromagnetic radiation produced with time by welding operations, and presented a report of his study with a portable instrument that had been developed in the USSR which integrates the total light output in the range of 280 to 400 nm. He found that the strength of the radiation varied with the angle from the plane of the welding arc.

1.5 Noise

In a study of noise produced by various welding processes, Hermanns (Ref. 65) reported that, with the exception of gas welding with a size 3 nozzle and GTAW, all processes studied (including gas welding, SMAW, GMAW shielded with inert gases, CO_2 or mixed gases, GTAW, and acetylene, propane, or plasma cutting) produced at least occasional noises above 85 dB (A) at a distance of 400 mm from the source (Table 3). With GMAW, the noise generated varied with the electrode used and more noise was produced than with SMAW. Of the welding processes studied, plasma cutting was the noisiest, and the noise level varied slightly with the metal being processed (plasma cutting of copper or aluminum was slightly noisier than that of steel). Hermanns recommended that reduction of the size of the flame be used to reduce noise levels during oxyacetylene welding.

Table 3
Noise levels of various welding processes*

Activity	Pulse sound level dB (A)
Gas welding, size 3 nozzle, 5 mm thick steel plate	84
Gas welding, size 4 nozzle, 5 mm thick steel plate	92
Gas welding, size 5 nozzle, 5 mm thick steel plate	97
Gas welding, size 6 nozzle, 5 mm thick steel plate	103
SMAW, 150A, AC, Ti VIII s rod electrode, core diameter 3.25 mm	86
SMAW, 180A, AC, Ti VIII s rod electrode, core diameter 4 mm	84
SMAW, 180A, AC, Kb IX s rod electrode, core diameter 4 mm	86
Slag chipping on plate (200 mm x 150 mm x 10 mm)	105
SMAW, 110A, core diameter 3.25 mm, pipe, including slag chipping	92
SMAW, 200A, core diameter 4 mm, fillet weld, including slag chipping	96
GMAW - mixed gas shield, 300A, wire diameter 1.2 mm, spray arc, steel	97
GMAW - CO_2 shield, 100A, wire diameter 0.8 mm, short arc	91...95
GMAW using a pulsed arc (100 Hz), 200/300A, wire diameter 1.6 mm, aluminum	95
GMAW, 200A, wire diameter 1.6 mm, aluminum	102
GTAW, 100A, rod diameter 2.4 mm, DC	65
GTAW, 60A, rod diameter 2.4 mm, AC	74
Flame cutting, acetylene, nozzle 10 to 25 mm, pipe, 6mm wall thickness	88
Flame cutting, propane, 10 mm plate	95
Plasma cutting, 100A, 10 mm steel plate	98
Plasma cutting, copper sheet	100
Plasma cutting, aluminum sheet	100
Plasma cutting, thick workpieces	>110
Arc-air gouging	>103
Grinding, manual grinder	105
Dressing, needle hammer	103

*Measured at a distance of approximately 400 mm from source

Hermanns, Ref. 65

2. Effects of Welding on Human Health

Potential health hazards are associated with all aspects of the welding exposures described in Section 1, including gases, vapors, fumes, noise, and radiation. Other hazards associated with welding are burns from hot flying metal pieces and the risk of electrocution (particularly for underwater welders). Welding exposures may directly affect the eyes (radiation, irritating gases, burns from flying metal particles), ears (noise, burns from flying metal particles), skin (radiation and burns), and musculature (strained working positions). With the exception of the latter, protective clothing and gear, when used properly, can effectively protect the welder. On the other hand, the potential health effects of welding emissions (gases, vapors and fumes) on the internal organs are more difficult to assess and control. The chronic effects of some of the individual components of welding emissions on humans are poorly understood and safe exposure levels cannot be determined with the available information. Also, little is known about the combined effects of many components of welding emissions. Some of these may interact synergistically, making their combined effects greater than, or different from, the additive effects of the individual components. Antagonistic effects are also possible, and the combined effects of two or more toxins could be less than their individual effects. These poorly understood properties of complex industrial emissions make it difficult to determine realistic permissible exposure levels for workers. For this reason, it is important for the industry to maintain an active surveillance of the health research literature and to remain well informed regarding any reports of adverse health effects of welding. Described below are health reports which appeared in the literature from December, 1980 (the time of writing the *Effects of Welding on Health*, Volume III) to June 1982.

2.1 Respiratory Tract

The respiratory tract represents the primary route of entry of welding fumes and gases into the body. Toxic components of the emissions may have either a direct effect on the lung or may be transported from the lung via the lymphatic and circulatory system to other parts of the body where their toxic effects may be manifested. Examples of gases that directly affect the lung are ozone and nitrogen dioxide which are highly irritating and may cause pulmonary edema after acute exposures. Emphysema and fibrosis may result from chronic exposures to these gases. Carbon monoxide has a systemic effect by combining with hemoglobin and interfering with the transport and utilization of oxygen. This will cause anoxia which can be fatal with sufficiently high exposures.

As described earlier, a large proportion of the particles in welding fume are respirable and may be deposited in the lung. Most of the particles which enter the lungs become ingested by macrophases within a day or two and will either be eliminated in this form by normal coughing mechanisms, or, for particles such as ferric oxide and alumina, may remain quiescent in the lung for long periods of time in pockets surrounded by these cells. Some particles may be transported from the deeper part of the lung to the lymphatic system. Relatively insoluble compounds, such as nickel oxide, may gradually become solubilized, leave the lung and lymphatic system, and become distributed to other parts of the body.

2.1.1 Pneumoconiosis

Pneumoconiosis is a general term for dust deposition in the lungs. Siderosis, or deposits of iron particles, is the most commonly occurring form in welders. Aluminum welders may have large deposits of aluminum particles in the lung. Both iron (Ref. 32) and aluminum particles (Ref. 110), in the absence of other foreign materials, are generally considered to be benign. Thus, they may remain in the lung for long periods without evidence of pathological changes in their vicinity. Fibrosis, the growth and spreading of connective tissues with resultant formation of scar tissue, results from the deposition of particles and fibers such as quartz (silica) and asbestos. This is a pathologic condition which, when extensive, can cause severe impairment of lung function.

The pathogenicity of ordinarily benign particles, such as those composed of aluminum, may change when they are trapped in the lung in conjunction with other materials. Bauxite pneumoconiosis, or Shaver's disease, a condition found in some aluminum smelters, is associated with inhaled particles of alumina and crystalline silica. The characteristic pathological lesion of Shaver's disease is a diffuse interstitial fibrosis as opposed to the focal or nodular type of fibrosis induced by silica alone.

It should be noted that most of the fibrotic particles referred to above (e.g., asbestos and silica) are not produced by welding and in those cases where they may be present in the welding environment, they arise from other processes.

2.1.2 Estimation of Retained Particles in the Lung

A magnetic method for the measurement of ferromagnetic particles in the lungs was developed by Cohen a decade ago (Refs. 29 and 30) and has been used by several investigative groups to estimate the retention of particles in the lungs of welders and persons in other dusty occupations. With this method, the subject's lungs are magnetized with an external magnetic field. Then the field is removed and the remanent field produced by the magnetized particles within the subject's upper torso is measured. In a recent study of asbestos miners and millers, Cohen and Crowther (Ref. 31) noted that those workers with welding experience had, on the average, five times more ferromagnetic particles in their lungs than did nonwelders.

Using this method, Freedman *et al.* (Ref. 42) found that the amount of ferromagnetic minerals retained in the lungs of eleven electric arc welders from the shipyard industry was several orders of magnitude greater than that present in three machinists from the same shipyard, 16 former asbestos insulators, and 24 rural male controls. The amount of materials retained in the lungs of welders correlated well with the number of years spent in the welding trade as well as the extent of siderosis indicated by chest X-rays, but not with smoking histories. The latter finding was unexpected by the investigators since an earlier report had shown that fewer particles are deposited deeply in the alveoli because of constriction of the smaller airways in the lungs of smokers (Ref. 110).

Kalliomaki *et al.* (Refs. 73 and 80) used magnetopneumography for estimating the amount of metal contaminants retained in the lungs of workers performing SMAW of mild steel, stainless steel welders engaged primarily in either GTAW or SMAW, iron factory and foundry workers, and stainless steel grinders. They found that stainless steel shielded metal arc welders had the highest mean quantity (approximately 4 grams of dust per person), and mild steel arc welders had the second highest mean quantity (approximately 1 gram of dust per person) of retained metals in the lung (Table 4). Gas tungsten arc welders had considerably less retained metals (0.2 gram per person) than the other welders. The magnetic moment and size of particles generated by GMAW, GTAW, and SMAW of stainless steel and SMAW of mild steel were examined, and it was found that the remanent magnetic field was indirectly related to the particle size. GMAW and GTAW generated the smallest particles which produced the greatest remanent field (Ref. 74).

Later, Kalliomaki *et al.* attempted to correlate levels of metals retained in the lungs with chromium and nickel levels in the urine of 83 SMAW and GTAW stainless steel welders (Ref. 75). Only the SMAW welders had elevated urinary chromium levels. Their urinary chromium levels were significantly related to the quantities of the metals in the lung. Urinary nickel levels were only slightly elevated in the shielded metal arc welders and were normal in the gas tungsten arc

Table 4
Groups, numbers, exposure time and the approximate amount of metal contaminants in the lungs of groups exposed to different metal aerosols

		Exposure time (years)		Amount of dust (G)	
	N	Mean	SD	Mean	Range
1. Mild steel arc welders from two shipyards	44	20	7	1.0	0.2...8.0
2. Stainless steel welders:					
GTAW (10% SMAW)	21	13	9	0.2	-
SMAW (90% SMAW)	29	19	10	4.0	0.6...10
3. Workers in an iron factory	27	13	5	0.2	0.02...1.0
4. Foundry workers	10	33	7	0.2	0.06...4.0
5. Stainless steel grinders	21	11	4	0.1	0.02...1.8

Kalliomaki *et al.*, Ref. 73

welders. The investigators attributed the absence of elevated urinary nickel levels in shielded metal arc welders to the solubility characteristics of the nickel fumes. However, as discussed below, other investigators observed elevated urine nickel levels in stainless steel welders (Refs. 47 and 114). Urinary nickel values may fluctuate markedly throughout the day, and 24 hour, or at least 8 hour, samples should be collected to obtain accurate values. In the study performed by Kalliomaki *et al.*, urine samples were collected only once for each person at the end of the work week.

2.1.3 Pulmonary Fibrosis

Two cases of pulmonary fibrosis in aluminum welders were recently reported. The first case involved a man, who had been an aluminum arc welder working primarily in confined spaces in the shipbuilding industry for 17 years (Ref. 134). This man was reportedly exposed exclusively to aluminum during his welding career and was also a tobacco smoker. He had reduced pulmonary function, and X-rays showed an area of the lung with a dense peripheral infiltrate suggestive of obstructive pulmonary disease. This part of the lung was surgically removed, and histological examination revealed diffuse and focal fibrosis with dense accumulations of macrophages containing metallic material, shown by scanning electron microscopy using EDXA to be composed primarily of aluminum. Pulmonary fibrosis is not generally associated with exposure to aluminum vapors, and the investigators do not exclude the possibility that other exposures, such as cigarette smoke, may have contributed to the pathogenic processes observed in the lung.

The second case was a 35 year-old aluminum welder whose clinical symptoms included moderate dyspnea (breathing difficulties) on exertion and decreased pulmonary function indicative of fibrosis (Ref. 62). Hazy basal infiltrates and a lung opacity were evident on chest X-rays. A biopsy showed consolidation in the opaque area and diffuse chronic interstitial pneumonitis. EDXA indicated that there were large amounts of aluminum in the lung. Small quantities of carbon dust, iron, and asbestos were also present. The authors concluded that the pulmonary fibrosis was largely due to the aluminum deposits since carbon and iron are known to be relatively inert in the lung and the small amounts of asbestos appeared to be "insufficient to account for his disease". However, as noted earlier, pulmonary aluminum deposits are generally benign unless they are associated with other toxic materials.

Gobbato *et al.* (Ref. 45) histopathologically examined the autopsied lungs from 17 individuals who had been electric arc welders in shipyards. Six died from nonpulmonary diseases, two from respiratory failure, eight from lung cancer, and one from mesothelioma. More than half the lungs had fibrotic lesions. Siderosis was observed in all of the lungs, sometimes in conjunction with infiltrates of inflammatory cells and sometimes in areas of fibrosis. The areas of inflammation and fibrosis always had associated iron deposits. These observations, as well as the large proportion of respiratory tract neoplasms, led the authors to suggest cautiously that there is a casual relationship between the pulmonary deposition of welding fumes and the development of advanced respiratory tract disease. These data should be interpreted with restraint since (1) the manner in which the cases were chosen for study was not indicated and any selectivity in this process could bias the results; (2) no history of tobacco use was obtained by the investigators; and (3) asbestos bodies were observed in three of the lungs.

Zober (Ref. 147) reviewed 29 reports published between 1955 and 1979 in which a total of 47 cases of histologically verified fibrotic changes in the lungs were documented. The mean age of the subjects was 51 years, with a range of 21 to 69 years. In 18 of the cases, pulmonary fibrosis was identified histologically during autopsy and in 29 cases, fibrosis was identified in biopsies or specimens that were collected surgically.

According to Zober, each of the authors of the 29 articles related the lung fibrosis to the welding exposure, usually because of the close proximity of fibrotic lesions and iron deposits. However, there was insufficient information concerning the history of tobacco use, previous medical history, extra-occupational and vocational exposures, and exposures to pollutants other than welding fumes at the workplace to categorically come to this conclusion. Among the 47 cases studied, 13 individuals had previous pulmonary diseases (e.g., tuberculosis or pneumonia), and in 14 cases there were indications of asbestosis or silicosis. Zober eliminated these and other cases in which there were no physiological symptoms of reduced lung function and concluded that, at the most, 20 cases remained in which welding fumes may have made a significant contribution to the development of lung fibrosis. Zober stressed the importance of performing studies which examine sufficient parameters using the appropriate methodology to enable comparisons between observations of different investigators at different occupational locations.

2.1.4 Acute Respiratory Infections

Because the inhalation of gases such as NO_2 reportedly reduces the resistance to respiratory tract infections, there has been some interest in whether welders have an increased frequency of acute respiratory tract diseases. Hayden *et al.* (Ref. 60) found that among 258 welders in the engineering industry, work absences due to upper respiratory tract infections were slightly, but significantly, greater than those of control workers. There was no difference in the frequency of absence resulting from other diseases between the two groups.

These results were only partially corroborated in a study by McMillan and Molyneux of workers in three shipyards in Devonport, Gibraltar (Ref. 84). Absence patterns of 533 welders were compared with two groups of controls. Group I was composed of men with intermittent exposure to welding fumes and consisted of 517 boilermakers and 835 shipwrights; Group II consisted of men who were not exposed to welding fumes and included 999 electrical fitters, 403 painters, and 765 joiners, all of whom were employed in the dockyards for at least 6 consecutive months during a recent 5 year period. Even though the mean age of the welders (44.8 years) was moderately higher than that of controls in Group I (37.5 years) and Group II (40.1 years), the absence rates were comparable among all three groups. After the data were adjusted for age, no substantial differences in the absence patterns from upper respiratory tract infections were seen. The number of episodes of lower respiratory tract disease was small in all groups and the proportion of workers in each group who were absent due to lower respiratory tract disease was comparable. However, welders and workers in Control Group I appeared to take longer to return to work after illnesses of this nature than those who worked in the cleaner environments (Control Group II).

The same workers were later evaluated in terms of absences due to respiratory tract illnesses and smoking habits (Ref. 83). It was found that a higher proportion of welders smoked than did controls. Among the nonsmokers in all groups, welders had the lowest frequency and severity of absences due to upper respiratory tract diseases. There was no difference between the absentee rates of controls and welders due to lower respiratory tract disease. Among the smokers, the severity of absence due to all respiratory tract diseases was similar in all three groups. In the case of lower respiratory tract disease, smoking welders appeared to have slightly more and slightly longer spells of absence than did controls. McMillan attributed this to their smoking rather than to exposures to welding fumes, stating that smoking welders are less likely to return to work before their symptoms are completely resolved since "they are more susceptible to the irritating and thus aggravating effects of welding pollutants during acute respiratory disease".

2.1.5 Lung Function Tests

Pulmonary function tests are used to detect disease processes which restrict lung expansion or reduce pulmonary elasticity. (The reader is referred to Appendix C of the *Effects of Welding on Health*, Volume I for more thorough explanation of the terms used in this section.) In brief, these tests measure, by spirometry, the air volume which can be inhaled or expelled either forcefully or under normal breathing conditions. An obstructive or restrictive ventilatory defect is a change in the ability to move air into and out of the lungs.

The forced vital capacity (FVC) is a measure of the volume of air that can be expelled forcefully following a maximal inspiration. This measure reflects resistance to airflow in the lungs. The forced expiratory volume during the first second of exhalation (FEV_1), a maximal mid-expiratory flow volume (MMFV), or maximal expiratory flow rate (MEFR) measure the percent of the total volume or the rate of air expelled during a poriton of the expiratory cycle. These measures all serve as indicators of airway resistance. Lung function tests are frequently used to monitor industrial workers for damage to the respiratory tract, such as changes in pulmonary compliance, and to detect early signs of pneumoconiosis or airway obstructions.

A factor that must be considered in interpreting these tests is the natural variation that occurs between individuals and with age. Another problem is that the smaller airways have little influence on the total airways resistance and, thus, lung function tests tend to detect only changes in the larger airways during dynamic compression of the lungs. The value of pulmonary function tests as a means of detecting pulmonary insufficiencies in welders has been controversial, and the results of tests by different investigators at different industrial locations are variable. At the very least, caution must be exercised in interpreting the results of lung function tests since they may not deviate from normal, even in the presence of small pulmonary lesions apparent on X-rays.

In a study of workers in four shipyards, Schneider and Rebohle examined pulmonary function in 432 electric arc welders and 420 control workers who were not exposed to welding fumes (Ref. 111). The welders and controls had similar work loads, were matched on the basis of age, and had no significant differences in smoking habits. The vital capacity, FEV, and mid-expiratory flow volumes and rates were the same for welders and nonwelders. The investigators concluded that, since welders are known to have a higher incidence of chronic bronchitis and pulmonary siderosis, the value of pulmonary function tests for medical screening of welders is questionable. However, data on the frequency of bronchitis and siderosis in the examined workers was not given.

Cavatorta *et al.* (Ref. 25) examined pulmonary function in welders who were exposed to relatively high levels of chromium through the use of chromium alloyed electrodes in an Italian machine shop for up to 2 years (average exposure to chromium fumes was 0.8 years). A total of 28 workers were examined who had been welders for a total of 2 to 40 years, with an average total work experience of 19 years. Although changes in respiratory function were related to the duration of time spent welding, a correlation between reduction of lung function and chromium exposure could not be established. As the investigators stated, this work may not have been conclusive because the exposure to fumes containing high levels of chromium was too short to have manifested toxic effects and also possibly because the total number of workers studied was insufficient for statistical analyses.

A comparison of the effects of welding and tobacco smoke on lung function was carried out with shipyard workers by Akbarkhanzadeh (Ref. 2). Pulmonary function tests were performed with 209 welders and 109 nonwelders. Approximately half of each group were cigarette smokers and the data were analyzed with respect to age, the extent of tobacco use, and the number of years spent welding. Tobacco smokers had a significant reduction of FEV^1, FVC, and T_l (the transfer factor in the lung for CO). Welding fumes also impaired lung function; this effect was independent of tobacco use. The effects of smoking and welding exposures on pulmonary function were approximately the same and no synergism was observed between the two exposures.

In a study performed in Japan (Ref. 97), exposure to ozone was found to have a definite effect upon pulmonary function. Flow volume curves were obtained from 68 healthy welders who had been exposed to relatively high levels of ozone while performing GMAW. The welding population was 20 to 29 years of age and was categorized according to the ozone concentrations in the areas in which they worked. Dust exposures and smoking habits were considered during the analysis of the data. Peak flow rate and maximum expiratory flow rate at 75% of the vital capacity (V_{75}) were decreased slightly in the welders. However, no correlation between these values and the ozone concentration to which the welders had been exposed could be made. On the other hand, the maximum expiratory flow rate at 25% vital capacity (V_{25}) of welders who had been exposed to ozone concentrations of 1.05 ppm or more was significantly lower than that of welders exposed to less than 1.05 ppm. The decrease in the V_{25} as well as the FVC was related to the length of exposure to ozone.

Hayden *et al.* (Ref. 60) found that welders in the engineering industry had no significant abnormalities in lung function. The values of FEV_1, PVC, peak expiratory flow, and maximum expiratory flow at 50% vital capacity ($MEFV_{50}$) were the same among 258 welders and 258 controls. The only differences were that the $MEFV_{25}$ was lower in all welders and the $MEFV_{50}$ was lower in welders who smoked tobacco products than in other groups. No correlation could be made between the results of lung function tests and the number of years spent welding, the work conditions (e.g., confined spaces, use of fume extraction), or the type of welding employed. These investigators also found that, during an experiment with nine welders, pulmonary function values were the same when measured before and after 15 minutes of continuous electric arc welding (Ref. 61).

Zober examined 40 welders and 40 age matched controls who were not exposed to welding fumes (Ref. 145). In each group, 18 of the subjects were smokers. The welders excreted four times as much chromium and nickel in the urine as did controls, confirming actual exposures to these substances. Acute, but not chronic, bronchitis was a more frequent complaint of welders than nonwelders. Among the nonsmokers, welders exhibited no increased abnormalities in pulmonary function. More smokers had abnormal forced expiration, FVC and $MEFV_{25-27}$ than did nonsmokers, and these abnormalities could not be related to the welding exposures. Of the smokers, six welders and two nonwelders had bronchitic murmurs whereas signs of bronchitis were seen in only two controls and in none of the non-smoking welders. The only significant pulmonary finding which was clearly related to welding exposures and not to tobacco use was siderosis, indicated by small roundish shadows on chest X-rays. Zober concluded that the "greater incidence of significant findings was clearly apparent in welders who smoked" and suggested that there was "probably a combined effect by tobacco smoke and welding fumes". The groups studied in this work were relatively small, however, and further studies using larger cohorts seem necessary before such conclusions can be substantiated.

It is apparent from these and from earlier studies (see *Effects of Welding on Health,* Volumes I, II and III) that inconsistent results may be obtained with pulmonary function tests. Stern (Ref. 122) discussed three factors which may confound the results of pulmonary function tests. The first is the effect of population dynamics, whereby self-selection among the welding population may encourage workers who experience ventilatory difficulties to change occupations. The second is the effect of smoking on pulmonary function. However, many investigators today are aware of this problem and attempt to control for the effect of cigarette smoke in the analysis of their data. The third factor is a bystander affect. Stern cites the example that many welders are employed in shipyards where there is known to be a higher incidence of chronic respiratory disease, among all workers than in other populations. Thus, he states that some of the pulmonary function test results may be related to exposures other than welding at the place of employment.

2.2 Cancer

2.2.1 Lung Cancer

A positive association between welding exposures and respiratory tract cancer has not been established although at least two components of some welding fumes, nickel and chromium, have been associated with respiratory tract cancer in workers from other industries. Certain nickel compounds have been implicated as causative factors in the development of nasal carcinomas and lung cancer in nickel refinery workers (Refs. 33 and 101). However, not all the specific nickel compounds involved in the development of this disease are known, and to date, evidence for an association between respiratory tract cancer and exposures to nickel-containing welding fumes has not been demonstrated. As with nickel, an association between respiratory tract cancer and exposure to chromium has been recognized for many years (Ref. 124). Only chromium in the hexavalant state (Cr-VI) has been implicated as an etiologic factor in lung cancer (Ref. 93). As with nickel, a clear cut association between exposure to chromium in welding fumes and the development of lung cancer has not been established. The identity of the Cr-VI compounds that may be responsible for the development of respiratory tract cancer is unknown, but the fact that the water-solubility of the Cr-VI may influence its bioavailability and hence its ability to cause cancer and other health problems has been the basis for regulatory decisions involving acceptable exposure limits (Ref. 91).

To explore the effects of exposure to nickel fumes, Polednak (Ref. 103) examined the causes of mortality of 1059 white male welders employed between 1943 and 1973 at three plants in Oak Ridge, Tennessee. Approximately half of these welders worked at the Gaseous Diffusion Plant where they welded nickel-alloy pipes and received high exposures to nickel fumes relative to the other cohorts who welded mild steel, stainless steel, and some aluminum. Standard mortality ratios (SMR) were calculated by comparing the number of actual deaths from different causes with those expected on the basis of death rates in sex- and age-matched controls in the U.S. population. Histories of tobacco use could be obtained only for about 30% of the study population.

The SMR's of all welders for deaths from all causes and for deaths from diseases of the respiratory tract and cancer did not significantly differ from those of the general U.S. population nor did they differ significantly when the two groups of welders were compared with each other. A nonstatistically significant excess of deaths from emphysema (SMR = 221) and lung cancer (SMR = 150) as well as a slightly elevated number of deaths from diseases of the nervous system (three deaths from brain cancer as compared with 1.55 expected and three from neurological diseases as compared with 1.89 expected) was observed when the entire cohort of welders was compared with the control population.

No correlation could be made between nickel exposure and respiratory tract disease when the two groups of welders were compared with each other. The effect of length of employment on the development of respiratory tract cancer was examined by comparing the SMR for welders with more than 50 weeks experience to those with less than 50 weeks experience (Table 5). For this comparison, the welders were placed into two categories; the first included all

Table 5
Mortality from selected causes by length-of-employment

Length of employment as a welder	Selected Causes of Death: All causes No.	SMR	All cancer No.	SMR	Respiratory cancer No.	SMR	Respiratory diseases No.	SMR
All welders								
<50 wk as welder (N=491)	100	97 (79;118)*	21	110 (68;168)	10	157 (75;289)	9	176 (81;334)
>50 wk as welder (N=568)	73	76 (60;96)	11	63 (31;113)	7	121 (49;249)	4	86
Nickel alloy welders								
<50 wk as welder (N=255)	44	84 (61;113)	9	94 (43;179)	2	62	3	116
>50 wk as welder (N=281)	40	80 (57;109)	8	90 (39;177)	5	175 (57;408)	2	83

*Confidence limits (95%) for SMR, if the observed number of deaths is five or more.

Polednak Ref. 103.

welders in the study, and the second consisted of the welders exposed to nickel fume. The only SMR that correlated with the duration of welding experience was that for respiratory tract cancer in the group of welders exposed to nickel fumes, but the confidence limits were high. Polednak concluded that the sample size and the length of follow-up were insufficient to assess thoroughly the risk of respiratory cancer in welders with nickel exposure. Because a small but insignificant excess of diseases of the respiratory tract and nervous system was observed among welders, Polednak concluded that follow-up studies of this and related groups of welders are warranted. This was supported by Gibson (Ref. 44) who noted that in Polednak's study, the "cut-off date for the ascertainment of vital status was January 1, 1974, and at least six additional years of follow-up are already available". In the welding population studied, 53% were hired before 1950 and had a 23 year follow-up period while another 40% were hired before 1960 and had a 13 year follow-up period. Cancer of the respiratory tract is generally thought to have a latency period of 20 to 30 years beyond the initial exposure to carcinogens. Therefore, a more conclusive analysis could be obtained by re-evaluating the causes of death among this population at a later time. The groups examined in this study are of particular value since welders from the same geographical location, with and without exposure to relatively high concentrations of nickel fumes, can be compared.

A cohort of 234 men who welded stainless steel for at least 5 years between 1950 and 1965 was selected from eight companies in Northern Sweden (Ref. 115). Their death rates from all causes did not differ from those of the general Swedish population. However, three welders died from lung cancer as opposed to 0.68 expected deaths ($p = 0.03$). Smoking histories were not obtained in this study, but the investigators cited other studies which indicated no important differences between smoking habits of welders and the general population. Sjogren suggested that the excess lung tumors may be related to exposure to hexavalent chromium and stressed the need for follow-up studies with larger populations of stainless steel workers.

Working on the hypothesis that the greater the exposure to pulmonary carcinogens, the earlier the age at which a person will succumb to lung cancer, Ahonen, using death certificates, determined the mean age of death of 1820 Finnish workers in different occupations (Ref. 1). The mean age of all cases of death from lung cancer was 65 years. Plumbers and insulators who, according to Ahonen have a greater than normal risk of lung cancer, had the most significantly depressed age of death from this disease. Welders and Smiths did not deviate from the average age of death from lung cancer, suggesting that their risk of dying from this disease is no greater or less than that of workers in other occupations.

Beaumont and Weiss (Ref. 16) examined the number of deaths from lung cancer among 3247 welders who were selected from the International Brotherhood of Boilermakers, Iron Ship Builders, Blacksmiths, Forgers, and Helpers in Seattle, Washington. These welders were part of a larger study for which the number of deaths from diverse causes was analyzed earlier (Ref. 17 and *Effects of Welding on Health,* Volume III). The welders worked in this occupation for at least 3 years between 1950 and 1976 and, for each, the initial day of exposure was before January, 1950. Of these welders, 529 had died by 1976. Their lung cancer rates were determined from death certificates and compared with age, sex, and race adjusted statistics for the total U.S. population and with the lung cancer death rates of 5432 nonwelders from the same union. Fifty of the welders died from lung cancer as opposed to 38 expected ($SMR = 132$, $p = 0.06$). The number of lung cancer deaths became statistically significant only when deaths occurring 20 or more years from the first date of employment were evaluated (observed: 39, expected: 22.4; $SMR = 174$, $p < 0.001$). No information concerning the smoking history of the study cohort was obtained.

A basic problem with health studies in the welding industry is that the occupational conditions are extremely heterogeneous; work conditions vary widely at different industrial locations, and the day-to-day variations in the working conditions at any one place make the occupational history of individuals difficult to document. Thus, it is important that the working conditions of study populations be carefully described. Zober (Ref. 146) examined 25 epidemiological studies of the chronic effects of welding on the respiratory tract which had been performed in 14 different countries between the years 1965 and 1980. He stated that variations in the design of the studies and insufficient background information made it difficult to compare them or to come to meaningful conclusions.

Some of the areas which Zober indicated were inadequately investigated or documented in many of the studies examined were: the descriptions of the welding processes used, dust levels in the breathing zones of the workers, duration of exposure, history of tobacco use, and the age of the welding and control populations. In many studies, the number of welders or the number of controls examined was too small to yield statistically meaningful results. Different parameters were examined in the various studies (e.g., chest X-rays were evaluated in eleven studies, clinical examinations were performed in three studies, examinations for bronchitis in one study, and pulmonary function tests in six studies). He concluded that the parameters studied and the analytical methods used should be standardized in order to make meaningful comparisons between welding populations possible.

In another critique of published investigations of

the respiratory tract disease incidence among welders, Anderson (Ref. 9) concluded that pulmonary diseases, including chronic bronchitis and cancer, are far more likely to be due to cigarette smoking and asbestos exposure than to welding fumes. In studies where welders in shipyards are compared with nonwelders in the general population, the data must be suspect because of the asbestos exposures that were frequently encountered in shipyards. Comparisons between age-matched workers in occupations other than welding within the same shipyard are more meaningful, because occupational exposures to contaminants other than welding fumes should be similar for both the welding and control populations. He concluded that "epidemiological studies must be controlled for smoking, for the use of drugs, for asbestos exposure, and for each industry in which the welder has worked, even if no welding was done in that industry". Prospective epidemiological studies should include quantitative measurements of the exposure to welding fumes, and care should be taken to control the habits of individual welders which may introduce variability into their total exposures.

Although these criticisms are valid, only a few of the requirements of these reviewers can be met in death certificate studies. Accurate data concerning smoking habits, exposure levels, extra-occupational and bystander exposures, physiological status, etc., can only be obtained with prospective studies; however, such investigations are enormously expensive. Death certificate studies, with their known limitations (e.g., a secondary cause of death may be listed on the death certificate; personal histories and habits are difficult to obtain), can provide an overall picture of the health status of a specific population and are useful for designating populations with environmental or occupational exposures in which health problems may exist and for which more intensive study is indicated.

Stern recently reviewed 19 published epidemiologic studies of welders. From the combined results of these studies, he calculated a small increase in the lung cancer risk (1.3 : 1) for welders in general (Ref. 120). After excluding the effects of tobacco smoking and shipyard employment where bystander exposures to asbestos and possibly other carcinogens frequently existed, Stern found that there was still a small increase in the cancer risk for welders. Arguing that chromium and nickel from welding fumes have the same toxicologic properties in *in vitro* and *in vivo* systems as do nickel and chromium salts, he discounted arguments that these elements in welding fumes present fewer risks than those in equivalent exposures in nonwelding industries. Carrying this argument further, Stern speculated that a large part of the excess lung cancer risk of welders could be attributed to the lung cancer rate in welders working with stainless steel and alloys containing nickel and chromium. Stern strongly emphasized the need to explore the possibility of "high-risk hot spots" within the industry. This would enable the limited resources of the welding community to be concentrated in areas where they are vitally needed (Refs. 120 and 122).

Stern's calculations are speculative and the elevation of the cancer risk which was derived from the 19 epidemiologic studies surveyed is quite low. However, the idea that "hot spots" may exist within the industry is a valuable concept, and the concentration of finances and research efforts in areas where there are exposures to suspect carcinogens or mutagens would seem to be indicated. One population that would serve well for this type of study is the nickel-alloy pipe welders in Oak Ridge, Tennessee, investigated by Polednak (Ref. 103). Another relevant study, currently being conducted by the German Cancer Research Center, will

Table 6
Mesothelioma rates in dockyard occupations

Occupation	Number of tumors	Man-years at risk	Rate per 100,000 man year
Laggers and sprayers	3	1,453	206
Boilermakers	6	3,720	161
Painters	4	3,366	119
Welders and burners	6	7,510	80
Shipwrights	11	17,890	61
Engine and electrical fitters	14	35,990	39
Joiners	1	5,140	19
Other intermittently exposed occupations	6	51,290	12
Unexposed workers	2	12,250	16

G. Sheers and R.M. Coles, Ref. 113

compare the cancer incidence in welders who worked with CrNi-alloyed filler metals with that in nonwelders in the same factory. The welding cohort of more than 1000 persons and a control group of similar size will come from over 20 different companies (Ref. 54).

2.2.2 Asbestos and Mesothelioma

The development of pleural mesotheliomas is closely associated with asbestos exposure. In a recent study, Sheers and Coles evaluated the incidence of mesothelioma that occurred between the years 1964 and 1978 among shipyard workers in Plymouth, England (Ref. 113). The mesothelioma rate of welders was found to be 80/100,000 man years as compared with 206 for laggers and sprayers, 161 for boilermakers, 119 for painters, and 16/100,000 man years for unexposed workers (Table 6). The time between the initial exposure and death from mesothelioma was 30 to 50 years. Thus, people developing mesotheliomas during the period of this study may have received the bulk of their exposure between 1945 and 1965, a time when the use of asbestos as an insulating material became widely accelerated, but the dangers of exposure to this material were not known.

Asbestos fibers were found to be in all locations in the shipyard used in this investigation. The concentrations varied widely in different parts of the shipyard at different times during the workshift. The fiber counts, which reached peaks in confined spaces such as engine and boiler rooms, were highest during the removal of lagging. The welders moved freely between work areas, and much of their exposure to asbestos apparently resulted from the general distribution of the fibers throughout the shipyard. Other exposures may have resulted from the use of asbestos cloth, gloves, and other materials. Since the results of this study reflect exposures that occurred some 40 years ago, at a time when there was little understanding of asbestos toxicity and the necessity for stringent precautions against asbestos exposure, the frequency of mesothelioma observed in welders and other workers may not be predictive of the future incidence of this disease.

2.3 Effects on the Ear and Hearing

Welders may suffer hearing loss from noises generated during welding as well as by other processes in the work area. For example, at a distance of 400 mm from the source, the mean pulse sound levels from manual grinding, dressing with a needle hammer, and slag chipping are approximately 105 dB. Noises generated by GMAW, SMAW, and gas welding are in the range of 85 to 95 dB (A) (Table 2) (Ref. 65). The OSHA noise standard is 90 dB (A), averaged over an 8 hour day.

Erlandsson *et al.* (Ref. 38) compared the effectiveness of earmuffs and earplugs in reducing hearing damage to welders and platers in an assembly and a boiler shop. No difference was seen between wearers of the different protective devices in the boiler shop, whereas assembly workers who used earmuffs suffered more severe hearing impairment and had a higher rate of progress of the hearing impairment than did users of earplugs. The investigators concluded that, although earmuffs reputedly afford somewhat better protection than earplugs, users of the latter may have less hearing damage because earplugs are less convenient to put on or remove and thus may be worn more constantly than earmuffs.

2.4 Effects on the Eye and Vision

In the absence of adequate protection, welders may suffer eye damage from exposure to nonionizing electromagentic radiation from the ultraviolet to the infrared regions.

Ultraviolet radiation is primarily absorbed by the outer structures of the eye (cornea, conjunctiva, iris). Exposure to ultraviolet radiation from the welding arc can result in acute keratoconjunctivitis, also known as arc eye, welder's flash, and actinic ray photokeratitis. The symptoms of this inflammatory disorder appear within 4 to 12 hours after exposure, last for up to 2 days, and include blurred vision, lacrimation, burning pain, and headache. A chronic photoophthalmia may also occur which is manifested by functional visual disturbances, increased light sensitivity, and symptoms of chronic inflammation of the conjunctiva and eyelids.

Visible radiation penetrates the eye and is absorbed by the retina and choroid. Thermal retinal damage can result from short, intense exposure to visible and near infrared radiation between 400 and 1400 nm. Blue light between 400 and 500 nm can cause a photochemical retinal injury (e.g., solar eclipse burn) referred to as the "blue light hazard". It is caused by an exposure lasting several seconds or longer to sources that are not sufficiently intense to cause retinal burns (Refs. 56 and 87).

Infrared radiation with wavelengths longer than 1400 nm is absorbed primarily by the cornea and aqueous humor where it may cause thermal damage. Infrared radiation of less than 1400 nm can be transmitted to the deeper structures of the eye and has been associated with the formation of lenticular cataracts.

In 1978, welding and cutting injuries represented 0.5% of the Workman's Compensation cases in the United States (Ref. 136). The results of a Bureau of Labor Statistic's survey showed that, of these injuries, 67% involved the eye and 1/3 of these were welder's flash. According to the welders surveyed, about 2/3 of flash burns were due to exposure to other welder's equipment. Most of the remaining ocular injuries were due to hot flying metal or slag particles. About half of the workers were wearing goggles or helmets with filter lenses at the time of their accidents. However, protective curtains or shields were used only at 25% of the work stations.

Emmett *et al.* (Ref. 36) found virtually no difference

in the frequency of eye abnormalities among 77 persons who worked as welders for at least 10 years, 75 nonwelders who worked in the welding area for at least 8 years, and 58 persons who worked for 10 years in the same fabrication facility but were never in the welding area. Visual acuity was the same in all three groups. The only statistically significant difference between the groups revealed by ophthalmologic examinations was the prevalence of dust in the eye lids. All persons in the welding shop area in this plant were required to wear safety glasses so these results may not be representative of shops where hygienic practices are less adequate. Despite the good practices, many of the welders reported having experienced flash burns. The frequency of flash burns that resulted from accidentally striking their own arcs was about the same as that from inadvertant exposure to other welder's arcs.

The case of a 56 year-old man who was formerly a welder and had a recurrent pterygium (an abnormal triangular fold of membrane, extending from the conjuctiva to the cornea) in one eye was recently reported (Ref. 68). The development of pterygia is apparently related to exposure to ultraviolet radiation (Refs. 126 and 127). However, welders are not known to have an increased incidence of this disorder (Refs. 36 and 96), a fact which has been attributed to the use of protective shields (Refs. 36 and 126). Work-related exposure to ultraviolet radiation was considered to be an unlikely cause of the pterygium case cited above (Refs. 96 and 126).

2.5 Effects on the Skin

Welders are subject to a variety of skin conditions, the majority of which can be prevented by protective clothing. Among these are burns from contact with hot metals, ultraviolet-induced erythema, and embedded airborne metal particles. In some individuals, allergic dermatoses can develop from repeated skin contact with components of welding fumes such as chromium, nickel, zinc, cobalt, cadmium, molybdenum, and tungsten compounds. As evidenced below, burns represent the most frequent skin lesions in these workers.

In a recent study in the Soviet Union (Ref. 131), 117 welders in a shipyard were examined for skin lesions. Work-related lesions were found in 45% of the welders. Of these, superficial and deep burns were the most common and were present in 41% of the workers. Ultraviolet radiation-induced dermatitis was detected on the face, hands, and forearms of 8.3% of the workers.

Burns were found to be more prevalent among welders than workers in other occupations in the Odense district of Denmark. Of 918 occupational skin burns and 948 ocular burns treated in the Odense Hospital, nearly half occurred in the iron industry, particularly during welding, cutting, and flame planing (Ref. 112).

Emmett *et al.* (Ref. 36) compared the frequency of skin abnormalities among 77 welders (Group 1), 75 nonwelders who worked in areas where they were exposed to welding fumes (Group 2), and 58 persons who worked in the same plant under conditions in which they were not exposed to welding fumes (Group 3). This was the same population that was used in the eye study cited above. The principle welding processes used were GMAW shielded predominantly with CO_2 and FCAW of mild steel. Occupational and smoking histories and details of sun exposures were obtained for each participant in the study. To control for variations in susceptibility to ultraviolet radiation, only Caucasian males were included in the study.

No differences were observed between the incidence of skin tumors or premalignant lesions among the groups. Erythema, when present, was generally localized and confined to unprotected areas of the body. Erythema of the neck was considerably more frequent in welders. The incidences of facial erythema and actinic elastosis (degeneration of elastic tissue of the dermis resulting from exposure to ultraviolet radiation) did not differ significantly between occupational groups.

Small cutaneous scars, resulting from thermal and mechanical injuries, were most prevalent in welders and machinists (Table 7). No significant differences in the frequency of various other dermatoses were observed. The investigators noted that, although premalignant and malignant skin tumors could not be related to welding exposures, good hygienic practices were employed in this plant, and a relatively young work population was examined. Further studies are needed in which ultraviolet-induced diseases and premalignant and malignant skin conditions are examined in an older population of welders in which the duration of exposure and period of time after the initiation of exposure are sufficiently long to enable the expression of skin diseases that may result from chronic exposure to ultraviolet radiation.

Table 7
Percentage of workers with small cutaneous scars

	Group 1	Group 2	Group 3
Number of Small Scars			
0	7%	76%	65%
1 - 5	29%	15%	16%
>5	64%	9%	19%

Emmett *et al.*, Ref. 36

2.6 Effects on the Cardiovascular System

Heart rates during simulated welding in a laboratory setting were measured while subjects were standing and performing overhead welding, stooping in a downhand welding position, or crouching with the welding table tilted 30° from the horizontal (Ref. 130). The apparatus which was used to simulate welding consisted of a table with a V-shaped slit across the surface. The weld site was represented by a reflecting photoelectric cell disc. A light bulb at the end of the tip of a metal rod (the "electrode") represented the arc. Welding was simulated by keeping the bulb in contact with the photoelectric cell disc as it was automatically drawn down the length of the slit without resting the electrode on the table edge. The experimental situation was designed so that procedures normally practiced during welding, such as changing the electrode and striking the arc, were performed. The heart rate variation, as measured by electrocardiography, was greatest when the welder altered his posture to change electrodes. The heart rate was lower when welding than during the change of electrodes. The variation in the heart rate occurring between welding and electrode changes was greatest when the downhand position was used.

Hayden *et al.* (Ref. 61) found that there was no significant change in carboxyhemoglobin levels in nine subjects before and after 15 minutes of continuous electric arc welding. Breathing zone levels of carbon monoxide were measured during the welding period and were found to be low relative to the threshold limit value.

2.7 Effects on the Nervous System

2.7.1 Manganese Poisoning

Chronic manganese poisoning is a progressive disease with neurological and psychological manifestations. The susceptibility to manganese poisoning varies widely; some highly resistant persons may show no signs of disease after years of exposure while others may become ill after as little as 2 months of exposure (Ref. 92).

Early symptoms of manganese poisoning include apathy, anorexia, headache, spasm, weakness, and irritability. Signs of psychoses may develop including euphoria, impulsive behavior, absent-mindedness, confusion, and aggression. The disease can be arrested, and may be reversible at this stage, if exposure is stopped. Otherwise, the disease will continue to progress; disturbances in speech, gait and balance may appear; and tremors may develop with eventual serious neurological involvement. Advanced stages of chronic manganese poisoning are readily confused with Parkinson's disease.

Chandra *et al.* (Ref. 27) measured the manganese content of the urine, signs of neurological injury, and serum calcium in 60 welders, 20 each from three different plants. In Plant A, the work area was well ventilated and the welding electrodes contained 2.1% manganese. In Plant B, the work areas were partially ventilated and the welding electrodes contained 0.55% manganese. In Plant C, the welders worked in confined spaces, and the welding electrodes contained 0.45% manganese.

Positive neurological signs (tremors and deep brisk tendon reflexes in the extremities) were seen in 5, 10, and 9 of the welders in Plants A, B and C, respectively. The urine manganese and serum calcium levels were higher than those of the controls in all workers with positive neurological signs. As expected, the presence of neurological signs was not related to the duration of exposure to welding fumes. In both this and a previous study (Ref. 26), Chandra *et al.* reported that increased serum calcium may frequently accompany mild and moderate (but not advanced) stages of manganese poisoning. The investigators suggested that serum calcium levels may be of value in diagnosing early stages of manganese poisoning.

2.7.2 Vibration Disorders

The use of vibrating equipment such as glazing hammers and grinding machines may produce "white finger syndrome" (Raynaud's disease) or intermittant blanching and numbness of the fingers. Edling (Ref. 34) observed indications of this condition in 10 of 24 welders, all of whom had used vibrating equipment for at least 6 years.

A study of the prevalence of vibration damage in the fingers of 526 shipyard workers in the Soviet Union (including 202 ship assemblers, 185 electric arc welders, 46 gas cutters, and 93 painters) was performed by Yatsenko *et al.* (Ref. 142). The controls were 60 technicians and engineers from the same industrial location. The extent of damage was tested by measuring the threshold sensitivity of the middle fingers of both hands to frequencies of 63, 125, and 250 Hz.

A moderate (threshold sensitivity +15 to +25 dB) or large (threshold sensitivity equal to or greater than 25 dB) decrease in the sensitivity to vibration was observed in 13% of the ship assemblers, 10.8% of the electric arc welders, 4.3% of the gas cutters, 21.7% of the painters, and none of the controls. The extent of the impairment was related to the duration of exposure. In this shipyard, the welders frequently assisted the ship assemblers with their work which included the use of vibrating instruments. That the exposure to vibrating instruments was solely responsible for the decreased sensitivity to vibration could not be substantiated in this study since the painters, who did not use vibrating equipment, had a high incidence of decreased vibration sensitivity. On the other hand, the incidence of decreased vibration sensitivity correlated with the degree of exposure to dusts and fumes and with the prevalence of pulmonary disorders. The investigators speculated that a decreased sensitivity of the hands to vibrational stimuli is an early sign of white finger syndrome or the more advanced occupational sensory polyneuritis and suggested that vibration sensitivity measurements should be a part of routine medical examinations of workers in the shipbuilding industry.

2.8 Effects on the Liver

Alsbirk *et al.* (Ref. 6) evaluated liver function in stainless steel welders by measuring the levels of serum alanine transaminase and aspartate amino transferase. A small increase in these enzymes was observed which was most marked in welders who consumed significant daily quantities of alcohol.

2.9 Effects on the Musculoskeletal System

2.9.1 Shoulder Pain

Shoulder pain is the second most frequent cause of visits to orthopedic clinics (Ref. 57) and is a frequent complaint of industrial workers. Because there is some uncertainty about the correlation between the strenuousness of the work and the occurrence of shoulder pain, Herberts *et al.* (Ref. 63) conducted an epidemiologic study of shipyard workers in which the occurrence of shoulder pain in 131 welders was compared with that of 57 office clerks.

The prevalence ratio of supraspinatus tendinitis in welders was 18%, which was significantly higher than that of the controls. The number of years of welding experience and the rated level of shoulder muscle load could not be correlated with the development of supraspinatus tendinitis, but these factors could not be ruled out by this study.

Kadefors *et al.* previously reported that muscle fatigue and pain is greatest when welding in overhead positions (Ref. 72). In that study, muscle fatigue was measured by electromyography of the back and shoulder muscles. These same investigators have furthered their study of the fatigue of the muscles controlling shoulder movement with a new instrument, the spectral moment analyzer. This instrument measures the myoelectric activity in individual muscles and yields hard data upon which to base recommendations concerning work postures which would be least likely to be harmful.

Using this equipment, the myoelectric activity in shoulder muscles was measured in men who were holding 2 kg weights while simulating work positions often maintained by welders (Ref. 64). Eight basic postures were studied. With each, the elbow was bent in a 90° angle and the position maintained for one minute. Three working levels were investigated: the hand at the waist, the hand at shoulder level, and the hand in an overhead position. Each position caused localized fatigue in all muscle studies, but the overhead position was the most potentially damaging to shoulder muscles. The investigators concluded that, when possible, work positions should be changed so that welding is done with the hand at shoulder or waist level. However, if welding must be done in the overhead position, the results indicated that the static loading on the supraspinatus muscle is significantly affected by elbow positioning, and potential damage can be reduced by assuming certain positions during overhead welding.

One approach to muscle strain and related disorders in welders (and other workers) is to provide them with periods of rest and exercise designed to overcome strain. The symptoms of a group of welders who suffered lumbosacral radiculitis (inflammation of the root of a spinal nerve) were treated by a program of exercises to be followed routinely during the work break which were aimed at reducing tension in those muscles most subject to static stress. The beneficial results were most evident during inclement weather, which normally aggravates this condition (Ref. 132).

2.9.2 Posture

Welders differ considerably in the postures they assume relative to the fumes. In one study, it was found that there was no statistical difference in dust levels in the breathing zones of different welders, regardless of whether the welder was standing erect, bent over his work, sitting, or varying his work positions (Ref. 133). However, several investigators have indicated that posture has a significant influence on the exposure to fumes and gases (Refs. 39, 53, and 104). This is related to the characteristics of the plume in which welding emissions rise from the arc. As the distance above the arc increases, upward velocity due to thermal uplift decreases and the diameter of the plume increases. According to Grosse-Wordemann and Stracke (Ref. 53), to minimize exposure to fumes, the horizontal distance of the welder's nose from the arc should be as great as possible and the vertical distance above the arc should be minimized. The welding postures affording the minimum exposure are, in order: the horizontal-vertical position; vertical, upward welding while seated; vertical, upward while standing; flat position-seated; flat position-standing. According to the authors, the latter two positions, while being the least advantageous due to fume exposure, are the most economical because they yield the largest weld pool (Ref. 53), and probably are the least stressful to the musculoskeletal system (see previous section). Thus, a compromise must be struck between economy and muscle strain on the one hand and fume exposure on the other, and in this case, fume extraction and use of an air curtain may be the best solution.

2.10 Effects on the Urogenital Tract

2.10.1 Fertility

Kandracova (Ref. 76) evaluated fertility disorders in a total of 4200 male workers that were referred to him by medical facilities in various factories in plants in Eastern Slovak. Among these men, there were 69 welders, 192 pipe fitters, and 57 car mechanics. The remainder, who served as controls, worked in other professions and occupations in which there was never any exposure to welding fumes.

The quantity, viscosity, color, and pH of semen and the number, quality, morphology, and motility of sperm were determined. The frequency of abnormalities was the same among all occupational groups, and it was concluded that the frequency of abnormal spermiogenesis among welders was the same as that among workers in other professions. These data are in accord with those previously reported by Haneke (Ref. 58).

2.10.2 Effects on Kidney Function

Alsbirk *et al.* (Ref. 6) detected no alterations in kidney function of stainless steel welders, as evaluated by the measurement of albumin and beta-2-microglobulin in the urine.

2.11 Effects on the Teeth and Oral Cavity

During a study of the condition of teeth and gums, Wulf and Seefeld (Ref. 140) observed that the incidence of periodontal disease was significantly greater in 100 electric arc welders than in 100 nonwelders (locksmiths, smiths, and cutters). The investigators speculated that particles in welding emissions could adhere to the teeth and gums, influencing the quantity and quality of dental plaque. The nitrogen oxides, following conversion to acids in the moist environment of the upper respiratory tract, could inhibit saliva production. The resultant dryness of the mouth could lead to inflammatory changes in the gingiva. Finally, the authors suggested that carbon monoxide may cause changes in the blood vessels in the mouth, producing gingival edema.

The observations of Wulf and Seefeld (Ref. 140) find support in the investigation of Melekhin and Agarkov (Ref. 89) who, during their study of fluorine and manganese sensitization in 247 electric arc welders (described below), also examined the condition of the oral cavity. They found that welders had 1.7 times more film deposit on their teeth and 3.1 times greater inflammation of the mucous membranes of the mouth (including periodontitis) than did the 100 control workers.

Another effect of welding on the oral cavity is the possible degeneration of dental amalgams in divers who perform underwater electric arc welding. To study the cause of complaints by divers of a metallic taste in their mouth and damage to their dental fillings, Rockert, Christensson, and Orthendal (Refs. 28 and 108) performed laboratory experiments in which the electric current through different types of amalgam was measured underwater at various distances from a number of types of welding electrode. They reported that the current through the amalgam was dependent on the surface area and the distance between the tooth and the welding electrode.

2.12 Metal Fume Fever

Metal fume fever may be caused by exposure to zinc, copper, magnesium, or other metal fumes, but it is most frequently associated with zinc fumes. The symptoms appear 4 to 12 hours after exposure and are heralded by a sweet or metallic taste in the mouth and dryness or irritation of the throat followed by coughing, shortness of breath, general malaise, nausea, muscle and joint pains, fever, and chills. The symptoms last for 24 to 48 hours and have no sequelae. A short-lived tolerance to metal fumes develops. This tolerance may be lost during a short absence from fume exposure, and the symptoms may recur on the return to work after a weekend or holiday; hence, the alternate name "Monday fever".

Nine cases of metal fume fever were recently reported in Mexican electric arc welders (Ref. 18). The symptoms, which appeared approximately 10 hours after exposure, included fever, dry cough, vomiting, and a metallic taste in the mouth. Abnormalities, four of which were suggestive of chronic bronchitis, were apparent in chest X-rays of five of the welders.

2.13 Sensitivity to Fume Components

Because the incidence of contact sensitivity to chromium is very high in Poland, a study was performed to attempt to determine the primary sources of exposure to this allergen (Ref. 109). Two hundred and fifty persons with dermatitis and positive patch tests to potassium chromate were questioned to identify the most probable sources of the exposures that initiated or exacerbated their sensitivity to chromate. Of the 250 subjects, 132 had contact with chromates in an occupational setting; 16 of these reported exposure to welding fumes. The study determined that tanned leather and matches are the two most frequent sources of chromate exposure for the general population. It is not known if sensitized welders developed their initial response to chromium through non-occupational sources such as matches or shoe leather, or whether they became sensitized to chromium through welding exposures.

A case of a welder with an unusually severe allergic response to chromium was recently reported by Zugarman (Ref. 148). The worker had a recurrent eczematous eruption involving the cheeks, neck, and ear which developed 3 1/2 years after he began welding chrome-plated steel desks and chairs. He had a strong positive reaction to a potassium chromate patch test, but did not react significantly to other allergens.

The eczema cleared up within 2 weeks after stopping work but recurred within 3 days after returning to the workplace. Neither sunscreens, barrier creams, topical corticosteroids, nor improved ventilation at his work station prevented the eruption. He became so sensitized to chromium that he could not work in any part of the plant without developing a cutaneous eruption.

The allergic sensitization of shielded metal arc welders to manganese and fluorine was studied in a pipe plant in the USSR (Ref. 89). The sensitivity was tested with an *in vitro* assay which measured the response of isolated white blood cells to manganese and fluorine. The sensitivity of 176 welders was compared with that of 100 controls who were not exposed to welding fumes. Not one of the controls was found to be sensitive to manganese or fluoride by this test whereas about half of the welders were sensitive to either manganese or fluorine, and 20 to 25% of them were responsive to both elements. The relationship of the results of this *in vitro* assay to the actual physiological response of the welders to manganese or fluorine exposures was apparently not investigated; hence, the significance of the results cannot be evaluated.

2.14 Biochemical Changes

Mikhail *et al.* (Ref. 90) evaluated serum protein and lipid levels, as well as alterations in blood hemoglobin and heme synthesis in 16 tank welders in Egypt who had worked at this trade for up to 22 years. The average blood lead levels were 42 μg/100 g in welders and 27.5 μg/100 g in 10 control workers who had not had occupational lead exposure. None of the welders had clinical signs of lead poisoning even though there were some changes in the diagnostic parameters examined. There was a significant decrease in the quantities of blood hemoglobin and an increase in the urine levels of delta-aminolevulinic acid, both of which are indicative of the interference with heme synthesis that typifies lead toxicity. Increased levels of the serum enzymes glutamic oxaloacetic transaminase, glutamic pyruvic transaminase, and lactic dehydrogenase were observed which suggested altered liver function. A significant increase in serum triglycerides and beta-lipoprotein concurrent with a decrease in the phospholipid/cholesterol ratio was also observed. The investigators suggested that this may indicate a premature development of atherosclerosis; however, this interpretation is highly speculative.

Grandjean and Kon determined blood lead levels in 59 welders, 67 workers in other departments, and 42 retirees from a repair in refitting shipyard (Ref. 46) where lead exposures resulted primarily from repair welding of steel parts coated with anti-corrosive paints. The average blood lead levels of the welders (39 μg/100 ml blood) was significantly higher than that of workers in other departments (26 μg/100 ml) and retirees 23 μg/100 ml). Among the welders, 53 of 59 had levels above 30 μg/100 ml (the authors considered 20 μg/100 ml to be the norm in the average working population) and 18 of these were above 40 μg/100 ml. In contrast, 19 or 67 workers in other departments had elevated blood lead levels and only two of these were above 40 μg/100 ml. Blood lead levels in all retirees were less than 30 μg/100 ml. They gradually returned to normal (20 μg/100 ml or less) within 4 years of retirement in the nonwelders and 7 to 8 years in welders.

Grandjean *et al.* also measured nickel concentrations in body fluids of welders and other workers in the same repair shipyard as well as in a construction shipyard (Ref. 47). The control population consisted of 15 hospital personnel and 28 retired workers from the repair shipyard. Fitters and painters had higher plasma nickel levels than welders and riggers at both shipyards. Only the welders at the repair shipyard had elevated plasma nickel levels. In this group, 9 of 38 had significantly elevated levels as compared with 1 of 43 controls, and 4 of 24 welders at the construction shipyard. Of the welders, 3 had plasma nickel levels higher than 10 μg/liter which is considered to be a critical concentration (Ref. 67).

Urine nickel levels were examined only in workers from the construction shipyard. In this case, 9 of 24 welders, but only 3 of 13 painters, had nickel levels that exceeded control values. This contrasts with the observation that plasma nickel levels were higher in painters. As the authors pointed out, urine nickel values fluctuate markedly during the day. Therefore, the use of random samples in this study, rather than urine samples collected from a full 8 or 24 hour period, may have rendered the urine nickel values less accurate than the plasma values.

Suzuki *et al.* (Ref. 125) studied the relationship between urine levels of cadmium, copper, and zinc in welders using cadmium-containing silver solder. Their results suggested that cadmium accumulation in the body affects the excretion of copper and zinc.

Sjogren (Ref. 114) compared urinary fluorine, chromium, and nickel levels in welders and nonwelding industrial workers. Welders using basic electrodes had higher fluorine levels, and those welding stainless steel had higher chromium and nickel levels in the urine at the end of the day than did controls. There was a linear relationship between the breathing zone and urine fluorine concentrations in the welders. Stainless steel welders had higher urinary chromium and nickel levels at the end of the work day than did nonwelders in the metal industry. Unlike fluorine, urinary chromium levels in stainless steel workers increased throughout the work week even though airborne levels did not so vary. The relationship between the chromium concentration in the air and urine samples collected over the whole week was linear. However, the deviation was great and Sjogren concluded that urinary chromium levels can only be used to approximate airborne exposures. The relationship between airborne and urinary nickel level was not linear.

Sjogren's results are in general agreement with earlier reports (see *Effects of Welding on Health*, Volume I) and they suggest that of the three elements examined, fluorine, and to a lesser extent chromium, but not nickel, hold promise for quantitative biological monitoring of the exposure of welders to fumes.

Work in the area of monitoring personal exposure by measuring accumulation of metals in biological tissues is in progress at the Lund Institute of Technology in Sweden. The relationship between personal exposure and levels of different metals in parts of the human body, e.g., body fluids, hair, nails, etc., is being studied (Ref. 85). Personal exposure levels will be monitored by lightweight samplers, also developed at that institute, which allow detection of elements at levels well below existing TLV's in samples automatically collected during short, sequential time intervals (Ref. 20).

2.15 Human Fatality

The death of a 34 year-old welder, apparently due to inhalation of cadmium fumes, was reported in 1981 (Ref. 82). This man had been working for about 30 minutes with an oxyacetylene torch and silver solder. The solder, which contained over 20% cadmium, was not labeled as such by the manufacturer. Although the building was "large and airy with a high roof", the immediate work area was not well ventilated. Within hours after exposure, the welder developed a persistent cough and had difficulty breathing. His urine cadmium levels were 20-fold higher than normally acceptable levels. He died 5 days later, and acute sterile pneumonitis, typical of cadmium poisoning, was found at autopsy.

2.16 Occupational Medicine

In the past, workers in occupational medicine have stressed the need for pre-employment examinations to screen from work in this industry those who may have health conditions that would render them particularly susceptible to the potential health effects of welding. Equally important are periodic checkups of currently employed welders to ensure that their health status remains unaltered by their employment, and that they have not developed any health conditions that would predispose them to sensitivity to welding exposures. The assessment of the health status of prospective and active welders was recently re-emphasized by Spelbrink (Refs. 118 and 119). Because the respiratory tract is the major route of entry for welding emissions into the body and is also a major target organ for the potential toxic effects of welding emissions, health examinations focus on the respiratory tract but also examine other organ systems. Spelbrink recommended that pre-employment examinations include sputum analyses, pulmonary function tests, chest X-rays, and hearing tests, and that information concerning smoking and medical histories be obtained. Employment in the welding trade should be discouraged for those with a history of bronchial asthma, allergic conditions of the upper respiratory tract, chronic bronchitis, tuberculosis, or high blood pressure which does not respond to treatment. The author further recommended that pulmonary function tests and chest X-rays be obtained periodically after employment has begun and after any extended sickness or period of disability.

3. Toxicologic Investigations in Animals and in Cell Cultures

3.1 Animal Studies

3.1.1 Pulmonary Effects

When particles or other foreign materials enter the lungs, they frequently elicit an inflammatory response which is marked by increased surfactant levels and the migration of leukocytes (macrophages and polymorphonuclear leukocytes) into the lung. As part of the natural line of defense against foreign bodies, levels of certain hydrolytic enzymes become elevated within leukocytes and in their immediate surroundings. Although this response is usually short-lived, certain foreign substances, such as some bacteria or quartz particles, may elicit a chronic inflammatory reaction. Some investigators believe that leukocytes, actively responding to inflammatory substances, release biological mediators which stimulate the division of fibroblasts (connective tissue cells) and that the extensive stimulation of fibroblasts resulting from a continuous assault with highly inflammatory foreign particles is responsible for pulmonary fibrosis.

White *et al.* (Ref. 138) examined the acute pulmonary inflammatory response in rats to three chemically distinct welding fume particulates. Two fumes were generated by SMAW of mild steel with either a basic or rutile-coated electrode; the third fume was generated by SMAW of stainless steel with a rutile-coated electrode (Arosta 316L - 18% Cr, 11% Ni). The results were compared with those obtained with 99% pure titanium dioxide. One week after the particles were introduced into the lung by intratracheal instillation (a technique whereby materials are injected directly into the lung through a cannula which is passed into the pharynx and extends through most of the length of the trachea), the relative lung weight, the number of leukocytes (free cells) in the lung, and the quantity of surfactant and hydrolytic enzymes present in the lung were elevated. Each of these parameters was elevated by all three types of welding particles; the particles from the stainless steel system had a slightly greater effect than the other particles

Table 8
Body weight/lung weight ratio, free cell numbers and levels of pulmonary surfactant in rats 1 week after instillation of TiO_2 or SMAW particulates

Parameter studied	Particulate		Dose of compound used (mg) 0	0.5	1.0	2.5	5.0
Lung/body weight ratio x 10^4	TiO_2	(F)	68	63	72	--	81
	S. steel	(M)	50	64*	71*	61*	72*
		(F)	70	72	89*	89*	93*
	Basic	(M)	50	59	56	68*	68*
		(F)	70	77	82	76	90*
	Rutile	(M)	50	57	66*	68*	67*
No. free cells x 10^{-6}/g lung	TiO_2	(F)	9.2	11.9	12.3	--	19.0*
	S. steel	(M)	11.0	9.0	12.4	18.7*	10.6
		(F)	8.0	11.8	14.2*	11.8	14.1*
	Basic	(M)	11.0	9.6	8.2	7.2	15.2
		(F)	8.0	19.5*	14.1*	16.2*	15.2*
	Rutile	(M)	11.0	12.4	9.3	9.4	14.6
Weight pulmonary surfactant (mg/g lung)	TiO_2	(F)	0.40	0.37	0.31	--	0.59
	S. steel	(M)	0.62	1.17	0.81	1.22	1.24*
		(F)	0.29	0.81*	1.08*	0.93*	0.98*
	Basic	(M)	0.62	0.92	1.12	0.74	1.33
		(F)	0.29	0.85*	0.70*	0.84*	0.93*
	Rutile	(M)	0.62	--	0.97	0.80	1.17

M = male; F = female

* = significant change from equivalent control group of rats

White *et al.* Ref. 138

(Table 8). The investigators speculated that the maximum response elicited by stainless steel particles results from the Cr-VI present in the samples.

In a follow-up study, White *et al.* (Ref. 137) attempted to determine the extent to which the inflammatory changes could be attributed to the water-soluble fraction of the particles which contains most of the Cr-VI and whether any of the observed pulmonary changes were persistent. The water-soluble and water-insoluble fractions of the stainless steel particles were instilled into the lung and the inflammatory effects compared with those resulting from the instillation of comparable quantities of potassium dichromate. The effects were studied over a period of 1, 4, and for the potassium dichromate only, 13 weeks after instillation of the test material. The inflammatory response to the instilled substances tended to decline after one week. The insoluble particles induced a greater increase in levels of hydrolytic enzymes whereas the number of leukocytes and the quantity of pulmonary surfactant were greater with the soluble fraction. Potassium dichromate enhanced most of the parameters examined. Although the welding fumes mimicked the effects of the potassium dichromate, the effects studied by these authors are the same as the general responses to many types of inflammatory stimuli, and hence it is not possible to conclude from this study that the inflammatory effects of welding particles in the lung are related to the Cr-VI content of the particles.

In a similar study (Ref. 7), rats and guinea pigs were exposed by inhalation and intratracheal instillation to flux-coated electrode fumes (the welding process from which the fumes were derived was not described). A normal inflammatory response was observed immediately after a single exposure to the fumes by inhalation. Continuous administration of welding fumes by intratracheal instillation eventually led to the appearance of fibrous tissue around aggregates of welding particles. Hydroxyproline was elevated in the lungs of animals exposed for a prolonged period. This amino acid is unique to collagen, the backbone material of connective tissue, and its elevation is indicative of an active fibrotic process. The investigators speculated that the fibrogenic response was due to silicon, which constitutes 7.7% (w/w) of the fumes studied. However, this is unlikely since, when it is present in welding fumes, silicon is generally amorphous and not in the quartz form (silica) associated with pulmonary fibrosis (Ref. 35).

Il'nitskaya and Kalina (Ref. 69) examined the effects in rats of inhaled mixtures of aluminum oxide (gamma-alumina) and electric arc welding gases. The concentrations administered were similar to those to which welders may be exposed. The first group of rats was exposed to aluminum oxide dust at a concentration of 48.6 mg/m^3 (particle size 8-11 μm); the second group was exposed to electric arc gases in which the ozone concentration was 1 mg/m^3, and the nitrogen oxides were present at 3.3 mg/m^3/. The third group exposed to a mixture of alumina dust and welding gases at the same concentrations administered to the first two groups. Animals in the fourth group were not exposed and served as controls.

By the end of the 4-month exposure period, minor changes in the hematologic profile and serum enzymes were seen in all of the experimental groups. Pulmonary changes occurred after 2- to 3-month exposures in all groups. The animals exposed to alumina dust (groups 1 and 3) developed moderate catarrhal bronchitis and swelling of alveolar septa with a small amount of perivascular edema. (The edema declined in group 3 towards the end of the observatin period.) The pulmonary effects observed in animals exposed to the gases alone were somewhat more severe than those in the other two groups. Bronchitis, "spreading" emphysema, swelling of the intra-alveolar septa, and perivascular edema were found in these animals. The investigators suggested that the adsorption of the electric arc gases by the alumina particles may have decreased the toxicity of the gases. Experiments such as these demonstrate the importance of evaluating the toxic effects of the complete mixture, rather than isolated components, of industrial or environmental substances to which man is exposed.

Havrankova and Skoda used the wet weight and DNA content of the lung as indicators of pathologic changes in the respiratory tract (Ref. 59) resulting from intratracheal instillation of fume samples from welding with three different basic electrodes and one rutile elecrode. All four samples caused an increase in the lung weight, and the fumes from the rutile electrode and one of the basic electrodes caused a statistically significant decrease in the DNA content of the lung.

The American Welding Society is currently planning a study which will compare the toxicity and fibrogenicity of standardized welding emissions generated by six processes. The emissions from three SMAW methods, two GMAW methods using either argon or CO_2 shielding gases, and one FCAW process using a mixed shielding gas of argon/CO_2 will be administered to rats by inhalation. The exposure levels will simulate those experienced during the welding in the workplace. The fibrogenicity of these emissions and toxic effects on alveolar macrophage function, in particular viability, will be compared. Studies of the mutagenicity of urine from exposed animals, cytogenetic effects in bone marrow, and systemic effects of the welding emissions are also being considered.

3.1.2 Whole Body Effects

Some question exists about the relationship between the water-solubility and the biological activity of the nickel compounds present in welding fumes. To investigate the influence of its water-solubility, English *et al.* (Ref. 37) studied the whole body distribution and excretion of soluble ($NiCl_2$) and insoluble nickel (NiO) in rats. The radiolabeled compounds were administered by intratracheal instillation and nickel levels in 14 organs, and feces and urine were determined at various times during the 90 days following treatment. $NiCl_2$ was readily distributed throughout the body and was rapidly cleared. About 70% of the administered dose was removed by the third day, and the main route of excretion was via the kidneys. (Ten times as much radioactivity was measured in the urine as in the feces over a 90 day period.)

NiO was more slowly distributed from the lungs to the rest of the body. Both the lungs and associated lymph nodes retained significant quantities of the NiO over the 90 day period whereas only negligible quantities of $NiCl^2$ were still present in those organs by the end of that time (Fig. 1). Whereas the $NiCl_2$ rapidly

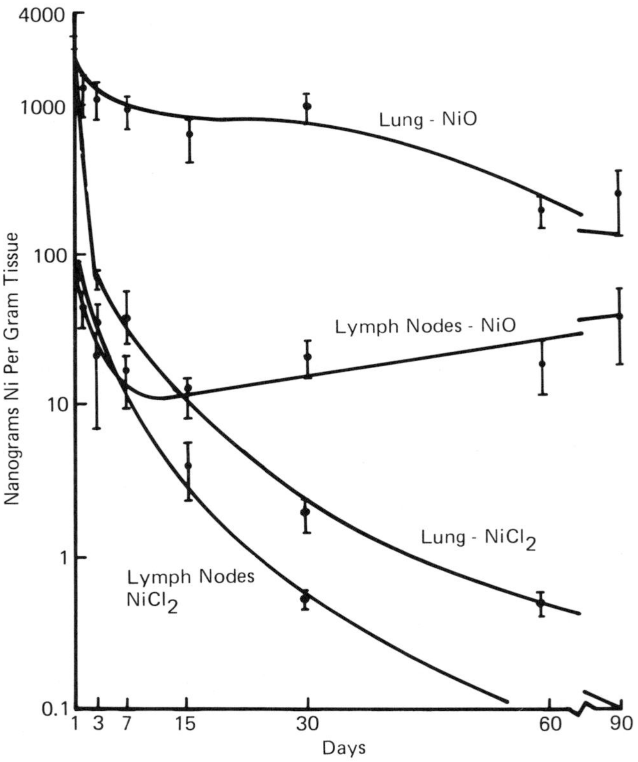

Fig. 1 — Change in nickel concentration in rat lung and associated lymph nodes following an intratracheal injection of $NiCl_2$ or NiO

English *et al.*, Ref. 37

entered and disappeared from the blood stream, the blood nickel levels remained almost constant during the first 2 months after instillation of NiO, suggesting that NiO was slowly and continuously cleared from the lung. Although the distribution rates were slower, significant levels of nickel were eventually found in other organs. Unlike $NiCl_2$ about half of the excreted NiO was present in the feces. Because the decline of nickel from all organs, with the exception of the lung, lymph nodes, and blood, was similar with the two forms of nickel, the investigators suggested that NiO may be removed from the lung by the slow conversion to a more water-soluble form whose behavior in the body is similar to that of $NiCl_2$. This concept is in accord with observations that the solubility of nickel in welding fumes is markedly increased by the addition of certain biological fluids (Ref. 128).

In a recent study, Kawata *et al.* examined the toxicity of basic-coated electrodes containing lithium salts (Ref. 77). (The use of lithium in electrode coatings is currently being explored as a means of reducing some of the health hazards in welding fumes; e.g., reduction of Cr-VI content of fumes - Ref. 128.) Animals exposed by inhalation to high levels of fumes from electrodes in which some of the potassium and sodium salts were replaced by lithium survived for 24 hours while those exposed to equivalent concentrations of fumes from electrodes without lithium died. The total quantity of sodium, potassium, and lithium oxides were lower in the fumes from electrodes containing lithium than in the other electrode tested. Therefore, it is possible that differences in the toxicity of the electrodes may have been related to differences in the alkalinity of the fumes. Of the tissues studied (lung, liver, kidney, and blood), the lung had the highest concentration of lithium after exposure to lithium-containing electrodes. The kidney, with the next highest level, had approximately one tenth the concentration found in lung. Fifty eight percent of the lithium in the kidney was excreted within 24 hours, and 93% was excreted by the end of one week. The investigators stated that further experiments are needed to determine the effects of lithium on kidney function.

Toxicologic studies of welding fumes are currently in progress at the University of Bradford in England (Ref. 48). In this work, fumes are being administered to animals by inhalation, intratracheal instillation, intramuscular injection, and skin painting.

3.1.3 Generation of Test Samples

The method by which samples are generated and prepared for biological studies is very important. When collecting and preparing samples, care should be taken that they are not physically or chemically denatured. Particles should not be allowed to aggregate or agglomerate if they are to be administered to animals by inhalation or intratracheal instillation. The method by which welding emissions are generated should be reproducible so that experimental data can be compared from laboratory to laboratory. For inhalation experiments, the time between generation of samples and delivery to animals should be the same as that between generation of fumes in the work situation and actual human exposures because of changes that occur naturally in air after the emissions form.

Recently, Gray *et al.* (Ref. 50) developed an apparatus for the reproducible generation of fumes from semiautomatic welding equipment which can be used for toxicological investigations. Fumes can be collected on filter pads and used later for *in vitro* investigation or for delivery to animals by methods such as intratracheal instillation. Alternatively, for animal exposures, the fumes can be delivered directly to inhalation chambers as they are generated. Palmer *et al.* (Ref. 100) developed a system for the collection of complex industrial emissions which are to be delivered to animals by intratracheal instillation. With this system, samples are collected directly into an aqueous medium and are ready for delivery to animals as they are removed from the sample collection device.

3.2 In Vitro Studies

In vitro assays of the toxicity of test substances are frequently performed in isolated cell or tissue cultures because they are quicker, easier to control, and generally less expensive than studies in live animals. Such tests can be used for routine screening of chemicals and complex mixtures such as welding emissions of unknown toxicity. Test substances that are positive in these assays can then be subjected to the more extensive and costly animal toxicity tests.

Assays of genotoxicity examine whether a substance causes mutations or alterations in the genetic material (DNA). This is important because some diseases in man, e.g., cancer, may be caused by such alterations in DNA. One such test of the genotoxicity of welding fumes was recently performed by Niebuhr *et al.* (Refs. 94 and 95). The method employed was the sister chromatid exchange assay which tests the ability of a chemical to induce certain types of chromosomal damage (e.g., the exchange of pieces of DNA strands between sister chromatids). The welding fumes tested in this assay were generated by argon-shielded GMAW of mild steel and cast iron using nickel-rich (95% nickel, 4% manganese) electrodes. The solubility of the fumes in water, which was extremely low (on the order of 0.1%), was increased (to about 1%) by the addition of fetal calf serum, which apparently converts the nickel salts to organic nickel complexes. The solubility of the fumes was markedly enhanced by the addition of whole blood, which, by its buffering capacity, maintains the pH at 7.4 and prevents the formation of the insoluble salt $Ni(OH)_2$ (Ref. 128).

The mutagenicity of the water-soluble and serum-soluble fractions of the fumes was tested independently and compared with that of water-soluble $NiSO_4$. The nickel fumes were found to be mutagenic in this test. The serum-soluble fraction of the fumes caused more chromosomal damage (per mol nickel) than did the water-soluble fractions derived either from the welding fumes or from nickel sulfate. The investigators concluded that an important influence on the genotoxicity of nickel in industrial fumes is the bioavailability of the nickel which may vary with the body fluids in which it is bathed. Because of this, they concluded that it would be inappropriate to use either the water-solubility or the total nickel content of fumes for the regulation of permissible exposures to industrial fumes (Ref. 95).

Other *in vitro* tests can examine the effects of chemicals on cell function or viability. Recently, White *et al.* (Ref. 139) compared the cytotoxicity of particles collected with 0.2 μm filters from SMAW using rutile-covered stainless steel electrodes and basic or rutile mild steel electrodes. Hexavalent chromium represented about 3.5% of the particles collected from the stainless steel system whereas only trace quantities of hexavalent chromium were detected in the samples collected from the other two electrodes. In the first test, saline-suspended particles were added to cultures of rapidly proliferating human cancer cells. Only the particles from the stainless steel electrodes inhibited proliferation. Since the dose-response curve obtained when hexavalent chromium was added to the cultures was almost superimposable on that obtained with the particles from the stainless steel system, the investigators conjectured that the toxic effects were due to hexavalent chromium.

In a second test, hemolysis (disruption of the red blood cell membrane with subsequent liberation of hemoglobin) of red blood cells by each of the particle samples was examined. In this case, dusts from the basic electrode were the most toxic. Finally, all three dusts caused about a 30% decrease in the viability of cultured macrophages. The investigators did not speculate about the factors responsible for the cytotoxicity to red blood cells and macrophages.

References

1. Ahonen, A. "Age at Death from Lung Cancer in Different Occupations." In *Abstracts Second World Conference on Lung Cancer,* June 9-13, 1980. Hansen, H.H. and Dombernowsky, P. (eds.) New York: Elsevier North-Holland, 1980.

2. Akbarkhanzadeh, F. "Combined Effects of Cigarette Smoking and Occupational Air Pollution in Welding Environment." Med. Lav., 72(1):61-6 (1981).

3. Alekseeva, I.S., Norkin, Y.I., and Andronik, V.A. "Hygienic Assessment of Electric Arc Welding of Copper with a Nonmelting Electrode in Nitrogen." Svar. Proizvod., (10):37-8 (1981).

4. Alekseeva, I.S., Norkin, Y.I., and Chumakova, I.V. "Hygienic Characteristics of the Air in the Manual Argon Arc Welding of Copper Alloys." Svar. Proizvod., (2):40-1 (1981).

5. Alekseeva, I.S., Ryzhikov, I.E., and Norkin, Y.I. "Hygienic Evaluation of Argon Arc-Welding with a Nonmelting Electrode Using Nickel-Containing Materials." Gig. Tr. Prof. Zabol., (5):56 (1981).

6. Alsbirk, K.E., Mogensen, C.E., Husted, S.E., and Geday, E. "Liver and Kidney Function in Stainless Steel Welders." Ugeskr. Laeg., 143(3):112-6 (1981).

7. Al-Shamma, K.J., Hewitt, P.J., and Hicks, R. "Effects of Flux Coated Electrode Welding Fume on the Lungs of Exposed Animals." 2nd Int. Congress on Toxicology, Brussels, Belgium July 6-11, 1980. Toxicol. Lett. (Amst.), (Special Issue 1):211 (1980).

8. American Welding Society. *Laboratory Method for Measuring Fume Generation Rates and Total Fume Emission of Welding and Allied Processes.* Miami, FL: American Welding Society, F1.2-79 (1979).

9. Anderson, R.S. "The Effects of Welding on Health: Evidence, Problems, Future Research Suggestions." Welding J., 61(4):40-4 (1982).

10. Andersson, H.C., Tranberg, E.A., Uggla, A.H. and Zetterberg, G. "Chromosomal Aberrations and Sister-Chromatid Exchanges in Lymphocytes of Men Occupationally Exposed to Styrene in a Plastic-Boat Factory." Mut Res., 73:387-401 (1980).

11. Andrews, L.R. and Hanlon, R.G. "Theoretical and Practical Considerations in the Analysis of Chromium in Welding and Other Metallurgical Fumes". Proc. 2nd Chromates Symp., pp. 244-58. Pittsburgh, PA: Industrial Health Foundation, 1981.

12. Arthaud-Richter, Y., Bresson, J.R., Chambon, P., and Prost, G. "Qualitative Study of Atmospheric Dust in a Welding Shop." Arch. Mal. Prof. Med. Trav. Secur. Soc., 41(1):33-4 (1980).

13. Barfoot, K.M., Mitchell, I.V., Verheyen, F., and Babeliowsky, T. "Combined PIXE and X-ray SEM Studies on Time-Resolved Deposits of Welding Shop Aerosols." Nucl. Instrum. Methods, 181(1-3):449-57 (1981).

14. Bartley, D.L., McKinnery, W.N., Jr., and Wiegand, K.R "Ultraviolet Emissions from the Arc Welding of Aluminum-Magnesium Alloys." Am. Ind. Hyg. Assoc. J., 42(1):23-31 (1981).

15. Baum, K. "Determination of the Ozone Contents of Tungsten - Inert Gas Welding with Analytical Equipment and Test Tubes - Comparison of Methods". Schweissen Schneiden, 30(11):439-41 (1978).

16. Beaumont, J.J., and Weiss, N.S. "Lung Cancer Among Welders." J. Occup. Med., 23(12):839-44 (1981).

17. Beaumont, J.J., and Weiss, N.S. "Mortality of Welders, Shipfitters and Other Metal Trades Workers in Boilermakers Local No. 104, AFL-CIO, USA." Am. J. Epidemiol., 112(6):775-86 (1980).

18. Bernal-Tapia, J.A., Sanchez Bernal, M., and Franco Trujillo, J. "Fever of Some Solderers. Information on New Cases." Rev. Med. Inst. Mex. Seguro Soc., 19(3):335-6 (1981).

19. Blakeley, St. J.H. and Zatka, V.J. "Determination of Hexavalent Chromium, Total Chromium, and Total Nickel in Welding Fume." INCO Metals Co.,

internal report (1978) as cited in Andrews and Hanlon (1981).

20. Bohgard, M., Malmqvist, K.G., Johnson, G.I. and Akelsson, K.R. "Personal Aerosol Sampler with Size Fractionation and Time Resolution." In *Aerosols in the Mining and Industrial Work Environment,* Vol. 3. Woburn, MA: Ann Arbor Science Publishers, 1982.

21. Bohme, D., and Heuser, H. "The Production of Air-Polluting Substances When Welding Over Production Coatings." Schweissen Schneiden, 34(2):72-77 (1982).

22. Buki, A.A., and Feldman, A.M. "Prediction of the Composition of Aerosol Formed in Welding in Shielding Gases." Weld. Pr. (Engl. Transl)., 27(2):8-12 (1980).

23. Carsey, T.P. "Feasibility Study of X-Ray Fluorescence for Analysis of Welding and Brazing Fumes." U.S. NTIS, PB82-116807.

24. Castagna, R., and Spagnoli, G. "Experimental Study on Hazards Related to Toxic Gas Emissions During Continuous Arc Welding in Argon Atmosphere." Securitas, 65(4):227-32 (1981).

25. Cavatorta, A., Falzoi, M., Mutti, A., Pedroni, C., Frigeri, G., and Franchini, I. "Spirometric Alterations in Chromium-Exposed Welders." Ateneo Parmense (Acta Bio-Med.), 51(4):299-304 (1980).

26. Chandra, S.V., Seth, P.K., and Mankeshwar, J.K. "Manganese Poisoning: Clinical and Biochemical Observations." Environ. Res., 7:374-80 (1974).

27. Chandra, S.V., Shukla, G.S., Srivastava, R.S., Singh, H., and Gupta, V.P. "An Exploratory Study of Manganese Exposure to Welders." Clin. Toxicol., 18(4):407-16 (1981).

28. Christensson, G., and Rockert, H. "Effects of Electrical Welding Under Water on Dental Amalgams and Enamel In-Vitro." IRCS Med. Sci.: Libr. Compend, 8(10):757-8 (1980).

29. Cohen, D. "Measurements of the Magnetic Fields Produced by the Human Heart, Brain and Lungs." IEEE Trans. Magn., MAG-11, No. 2, 694-700 (Mar. 1975).

30. Cohen, D. "Ferromagnetic Contamination in the Lungs and Other Organs of the Human Body." Science, 180:745-8 (1973).

31. Cohen, D., Crowther, T.S., Gibbs, G.W., and Becklake, M.R. "Magnetic Lung Measurements in Relation to Occupational Exposure in Asbestos Miners and Millers of Quebec Canada." Environ. Res., 26(2):535-50 (1981).

32. Creasia, D.A. and Nettesheim, P. "Respiratory Cocarcinogenesis Studies with Ferric Oxide: A Test Case of Current Experimental Methods." In Karbe, E. and Park, J.F. (eds.) *Experimental Lung Cancer: Carcinogenesis and Bioassays,* New York: Springer-Verlag, 1974.

33. Doll, R., Mathews, J.D., and Morgan, L.G. "Cancers of the Lung and Nasal Sinuses in Nickel Workers: A Reassessment of the Period of Risk." Brit. J. Ind. Med., 34:102-5 (1977).

34. Edling, C. "Welders' Health Examinations (letter)." J. Soc. Occup. Med., 30(2):86 (1980).

35. Eichhorn, F., Trosken, F., and Oldenburg, T. "The Production of Air Polluting Substances during Manual Arc Welding." Schweissen Schneiden, 34(2): 63-7 (1982).

36. Emmett, E.A., Buncher, R., Suskind, R.B., and Rowe, K.W., Jr. "Skin and Eye Diseases among Arc Welders and Those Exposed to Welding Operations." J. Occup. Med., 23(2):85-90 (1981).

37. English, J.C., Parker, R.D., Sharma, R.P., and Oberg, S.G. "Toxicokinetics of Nickel in Rats After Intratracheal Administration of Soluble and Insoluble Form." Am. Ind. Hyg. Assoc. J., 42(7):486-492 (1981).

38. Erlandsson, B., Hakanson, H., Ivarsson, A., and Nilsson, P. "The Difference in Protection Efficiency between Earplugs and Earmuffs." Scand. Audiol., 9(4):215-21 (1980).

39. Farwer, A. "Air Pollutants in Shielded Arc Welding; Relationship between Their Formation and the Concentration in the Breathing Zone - The Results of Laboratory and In-Plant Measurements." Schweissen Schneiden, 34(2):97-101 (1982).

40. Farwer, A. "Measurements of Ozone Concentrations during Inert Gas Tungsten Arc Welding of High Alloy Steel and Aluminum in a Fabrication Shop." Schweissen Schneiden, 32(11):437-442 (1980).

41. Frant, R. "Formation of Ozone in Gas Shielded Welding." Ann. Occup. Hyg., 6:113-25 (1963).

42. Freedman, A.P., Robinson, S.E., O'Leary, K., Goodman, L. and Stellman, J.M. "Non-invasive Magnetopneumographic Determination of Lung Dust Loads in Steel Arc Welders." Br. J. Ind. Med., 38(4):384-8 (1981).

43. Frei, A., Pirner, M., and Walser, H. "Measures against Ozone Formation in Arc Welding of Aluminum under Gas Protection." Schweisstechnik (Zurich), 70(10):215-20 (1980).

44. Gibson, N.E. "Mortality in Welders (letter)." Arch. Environ. Health, 37(1):60-1 (1982).

45. Gobbato, F., Melato, M., and Bucconi, S. "Welder's Lung of Arc Welders: Histopathologic Aspects and Possible Neoplastic Complications." Med. Lav., 71(2):132-40 (1980).

46. Grandjean, P., and Kon, S.H. "Lead Exposure of Welders and Bystanders in a Ship Repair Yard." Am. J. Ind. Med., 2(1):65-70 (1981).

47. Grandjean, P., Selikoff, I.J., Shen, S.K., and Sunderman, F.W., Jr. "Nickel Concentrations in Plasma and Urine of Shipyard Workers." Am. J. Ind. Med., 1(2)181-9 (1980).

48. Gray, C.N. University of Bradford, Bradford West Yorkshire, U.K. Letter to the author. Oct. 25, 1982.

49. Gray, C.N., and Dare, P.R.M. "Chromium in Metallic Aerosols." Abstract of Royal Society of Chemistry Papers from Conference on the Detection and Measurement of Hazardous Substances in the Atmosphere. City University, London. Dec. 1982.

50. Gray, C.N., Dare, P.R.M., and Hewitt, P.J. "Apparatus for the Controlled Generation of Fume from Semiautomatic Welding Processes." International Institute of Welding, IIW Document VIII 1019-82, 9pp. (Nov. 1981).

51. Gray, C.N., Hewitt, P.J., and Dare, P.R.M. "New Approach Would Help Control Weld Fumes at Source, Part One: Biomedical Background." Weld Metal Fabric., pp. 318-24, (Sept. 1982).

52. Gray, C.N., Hewitt, P.J., Hicks, R., and Blunsum, M. "Problems Involved in the Analysis of Chromium in Welding Fumes." International Environmental Safety Exhibition and Conference held at Wembley Conference Centre, London, Sept. 2-4, 1981, 8 pp.

53. Grosse-Wordemann, J., and Stracke, E. "Relationships between Exposure to Fumes When Welding and the Welding Position Dictated by Design Considerations." Schweissen Schneiden, 34(2):109-12 (1982).

54. Grothe, I. "Steps Taken by the West German Employers' Liability Insurance Associations to Protect the Health and Safety of Welders at Work." Schweissen Schneiden, 34(2):87-90 (1982).

55. Gvozdenko, L.A., Sokolov, M.V., Primak, V.N., Kuzina, A.S., and Shishkina, N.S. "Methods of Hygiene Research on Optical Radiation." Gig. Sanit, (2):51-22 (1980).

56. Ham, W.T., Mueller, H.A., and Sliney, D.H. "Retinal Sensitivity to Damage from Short Wavelength Light." Nature, 260(5447):153-5 (1976).

57. Hammond, G., Torgerson, W., Dotter, W., and Leach, R. "The Painful Shoulder Instructional Course Lectures. The Am. Acad. Orthopaedic Surgery 20:83-90. St. Louis, MO: C.V. Mosby Co., 1971.

58. Haneke, E. "Findings in the Ejaculate of Electro Welders." Dermatol. Monatsschr., 159(11):1036-40 (1973).

59. Havrankova, J., and Skoda, V. "DNA Content in the Lungs of Rats After Dust Application." Bratisl. Lek. Listy, 76(5):552-62 (1981).

60. Hayden, S P., Hayden, J., Pincock, A.C., Tyler, L.E., Cross, K.W. and Bishop, J.M. "Health of Welders in the Engineering Industry." 77th Annual Meet. Am. Lung Assoc. and 76th Annual Meet. Thoracic Soc., Detroit, MI, May 9-13, 1981. Am. Rev. Respir. Dis., 123(4 Part 2):125 (1981).

61. Hayden, S P., Pincock, A.C., and Bishop, J.M. "Pulmonary Function and Fume Concentration in Relation to a Brief Intensive Period of Electric Arc Welding." 77th Annual Meet. Am. Lung Assoc. and 76th Annual Meet. Thoracic Soc., Detroit, MI, May 9-13, 1981. Am. Rev. Respir. Dis., 123(4 Part 2):125 (1981).

62. Herbert, A., Sterling, G., Abraham, J., and Corrin, B. "Desquamative Interstatial Pneumonia in an Aluminum Welder." Human Pathol., 13(8):694-9 (1982).

63. Herberts, P., Kadefors, R., Andersson, G., and Petersen, I. "Shoulder Pain in Industry: An Epidemiological Study on Welders." Acta Orthop. Scand., 52(3):299-306 (1981).

64. Herberts, P., Kadefors, R., and Broman, H. "Arm Positioning in Manual Tasks: An Electromyographic Study of Localized Muscle Fatigue." Ergonomics, 23(7):655-65 (1980).

65. Hermanns, I. "Noise Problems When Welding--Causes, Effects and Prevention." Schweissen Schneiden 34(2):107-8 (1982).

66. Hinrichs, J.F. "The Welding Environment and Research." Weld. J. (Miami), 59(1):64,66 (1980).

67. Hogetveit, A.C., Barton, R.T., and Kostol, C.O. "Plasma Nickel as a Primary Index of Exposure in Nickel Refining." Ann. Occup. Hyg., 21:113-20 (1978).

68. Iliff, W.J. "Recurrent Pterygium in a Welder (letter)." J. Am. Med. Assoc., 246(9):1019 (1981).

69. Il'nitskaya, A.V., and Kalina, O.V. "Health Effects of Dust-Gas Mixtures in Plasma Technology." Gig. Tr. Prof. Zabol., (10):5-9 (1980).

70. Johansson, G.E., Malmqvist, K.G., Bohgard, M., and Akselsson, K.R. "Characteristics of Welding Fumes." Institutionen fur Karnfysik, Lunds Tekniska Hogskola, Lund, Sweden. Report LUTFD2/TFKF-3030, 32 pp. Aug. 21, 1981.

71. Johansson, M.I. "Gas Shielded Arc Welding and Working Environments." Svetsen, 39(6):5-8 (1980).

72. Kadefors, R., Petersen, I., and Herberts, P. "Muscular Reaction to Welding Work: An Electromyographic Investigation." Ergonomics, 19:543-58 (1976).

73. Kalliomaki, K., Aittoniemi, K., Kalliomaki, P.L., and Moilanen, M. "Measurement of Lung Retained Contaminants In-Vivo among Workers Exposed to Metal Aerosols." Am. Ind. Hyg. Assoc. J., 42(3):234-8 (1981).

74. Kalliomaki, P.L., Kalliomaki, K., Kelha, V., Sortti, V., and Korhonen, O. "Measurement of Lung Contamination among Mild Steel and Stainless Steel Welders." Soudage Tech. Connexes, 35(1-2):50-2 (1981).

75. Kalliomaki, P.L., Rahkonen, E., Vaaranen, V., Kalliomaki, K., and Aittoniemi, K. "Lung-Retained Contaminants, Urinary Chromium and Nickel among Stainless Steel Welders." Int. Arch. Occup. Environ. Health, 49(1):67-76 (1981).

76. Kandracova, E. "Fertility Disorders in Welders." Cesk. Dermatol., 56(5):342-5 (1981).

77. Kawata, K., Tanabe, Y., Adachi, T., Nagasaki, Y., Kobayashi, M., Maki, S., and Matsumoto, H. "Experimental Investigation of Hazards of Lithium in Welding Fume." Tekko Rodo Eisei, 29(3):22-4 (1980).

78. Kireev, V.I., Moslov, N.I., Roshchupkin, N.P., Schevchenko, S.A. and Matyazh, P.Y., "Hygienic Appraisal of an Aerosol Formed in the Welding of a Metal Coated with Protective Primers." Svar. Proizvod., (2):41-2 (1980).

79. Koerber, D., and Fissan, H.J. "Particle Emission from Welding of Painted Steel." Int. J. Environ. Anal. Chem., 10(1):13-21 (1981).

80. Koponen, M., Gustafsson, T., Kalliomaki, K., Kalliomaki, P.L., Moilanen, M., and Pyy, L. "Dusts in a Steel Making Plant: Lung Contamination among Iron Workers." Int. Arch. Occup. Environ. Health, 47(1):35-46 (1980).

81. Koponen, M., Gustafsson, T., Kalliomaki, P.L., and Pyy L. "Chromium and Nickel Aerosols in Stainless Steel Manufacturing, Grinding and Welding." Am. Ind. Hyg. Assoc. J., 42(8):596-601 (1981).

82. Lucas, P.A., Jariwalla, A.G., Jones, J.H., and Gough, J. "Fatal Cadmium Fume Inhalation." Lancet, 2(8187):205 (1980).

83. McMillan, G.H. "The Health of Welders in Naval Dockyards: Welding, Tobacco Smoking and Absence Attributed to Respiratory Diseases." J. Soc. Occup. Med., 31(3):112-8 (1981).

84. McMillan, G.H., and Molyneux, M.K. "The Health and Welders in Naval Dockyards: The Work Situation and Sickness Absence Patterns." J. Soc. Occup. Med., 31(2):43-60 (1981).

85. Malmqvist, K. Lund Institute of Technology, Lund, Sweden. Letter to the author. Dec. 1, 1982.

86. Malmqvist, K.G., Johansson, G.I., Bohgard, M., and Akselsson, K.R. "Elemental Concentrations in Airborne Particles from Welding and Metal Spraying Operations." Nucl. Instrum. Methods, 181(1-3):465-71 (1981).

87. Marshall, W.J., Sliney, D.H., Lyon, T.L., et al. "Evaluation of the Potential Retinal Hazards from Optical Radiation Generated by Electric Welding and Cutting Arcs." Nonionizing Radiation Protection Special Study 42-0312-77 (U.S.A. Environmental Hygiene Agency, Aberdeen Proving Ground, Md., 1977) U.S. NTIS, AD-A043023.

88. Mayer, A. and Salsi, S. "Chemical Pollution in Workplaces for Arc-Welding." Cah. Notes Doc., 101:485-97 (1980).

89. Melekhin, A.V. and Agarkov, V.I. "Sensitization of Fluorine and Manganese Compounds and the State of the Oral Cavity in Workers in the Electric Pipe Welding Industry." Gig. Tr. Prof. Zabol., (8):12-4 (1981).

90. Mikhail, T.H., El-Sawaf, H.A., Ibrahim, K.M., Awadallah, R., and El-Dessoukey, E.A., "Evaluation of the Effect of Lead Exposure on the Liver in Egyptian Lead Tank Welders." Z. Ernaehrungswiss., 19(1):50-6 (1980).

91. National Institute for Occupational Safety and Health. "Criteria for a Recommended Standard, Occupational Exposure to Chromium (VI)" U.S. Dept. of Health, Education and Welfare, Public Health Service, Center for Disease Control. DHEW Publ. No. (NIOSH) 76-129. U.S. NTIS, PB-248595 (1975).

92. National Institute for Occupational Safety and Health. "Occupational Diseases, A Guide to Their Recognition." U.S. Dept. of Health, Education and Welfare, Public Health Service, Center for Disease Control DHEW (NIOSH) Publ. No. 77-181:608pp. (June 1977).

93. National Research Council. Committee on Biologic Effects of Atmospheric Pollutants. *Chromium.* Washington, D.C.: National Academy of Sciences, 1974.

94. Niebuhr, E., Stern, R.M., Thomsen, E., and Wulf, H.C. "Genotoxicity of Nickel Welding Fumes Assayed With Sister Chromatid Exchange in Human Blood Cells in Culture." Ann. Clin. Lab. Sci., 11(1): 85-6 (1981).

95. Niebuhr, E., Stern, R.M., Thomsen, E., and Wulf, H.C. "Relative Solubility of Nickel Welding Fume Fractions and Their Genotoxicity in Sister Chromatid Exchange In Vitro." U.S. NTIS, PB82-158411, 13pp. (c1980).

96. Novak, J.F. "Recurrent Pterygium in a Welder (letter)." J. Am. Med. Assoc., 246(9):1019 (1981).

97. Ohmori, K., Tozawa, T., Ikemi, Y., Kobayashi, Y., and Kitazume, M. "Pulmonary Functions Observed by Means of Flow-Volume Curve in Welders Exposed To Ozone." Sangyo Igaku, 23(4):394-400 (1981).

98. Oleinchenko, K.A., Korneev, A.D., Zusin, V.Y., and Kabanets, A.N. "Effect of Ceramic Flux ZhA-64 Moisture on the Amount of Harmful Emissions during Automatic Aluminum Welding." Svar. Proizvod., (2):42-3 (1980).

99. Pabley, A.S., and Keeney, A.H. "Welding Processes and Ocular Hazards and Protection." Am. J. Ophthalmol., 92(1):77-84 (1981).

100. Palmer, W.G., Scholz, R.C., and Moorman, W.J. "An Innovative Approach to Sampling Complex Industrial Emissions for Use in Animal Toxicity Tests: Applications to Iron Casting Operations." Am. Ind. Hyg. Assoc. J., 44(3):184-189 (1983).

101. Pedersen, E., Hogetveit, A.C. and Anderson, A. "Cancer of Respiratory Organs Among Workers at a Nickel Refinery in Norway." Int. J. Cancer, 12:32-41 (1973).

102. Pokhodnya, I.K., Yavdoshchin, I.R., Bulat, A.V., and Shvachko, V.I. "Sources of Manganese and Iron Admission into a Welding Aerosol." Avtom. Svarka, (3):37-9 (1981).

103. Polednak, A.P. "Mortality Among Welders, Including a Group Exposed to Nickel Oxides." Arch. Environ. Health, 36(5):235-42 (1981).

104. Pomaska, H.U. "The Effects of the Welding Conditions and of the Workplace Conditions on the Air-Polluting Substances Produced During Shielded Arc Welding." Schweissen Schneiden, 34(2):102-7 (1982).

105. Press, H. "The Production of Air-Polluting Substances When Working with Fuel-Gases and Oxygen." Schweissen Schneiden, 34(2):68-72 (1982).

106. Press, H. "Formation of Nitric Oxides in Gas Welding. Measures for the Prevention of Injuries to Health." Soudage Tech. Connexes, 35(5-6):207-11 (1981).

107. Riediger, G., "Respiratory Protection When Welding." Schweissen Schneiden, 34(2):112-4 (1982).

108. Rockert, H.O.E., and Orthendal, T. "Effects of Electrical Welding Under Water on Dental Amalgams and Enamel." Undersea Biomed. Res., 8(1):64 (1981).

109. Rudzki, E., and Kozlowska, A. "Causes of Chromate Dermatitis in Poland." Contact Dermatitis, 6(3):191-6 (1980).

110. Sanchis, J., Dolovich, M., Chalmers, R., and Newhouse, M.T. "Regional Distribution and Lung Clearance Mechanisms in Smokers and Non-Smokers." In *Proc. of Symposium: Inhaled Particles and Vapours III*, Walton, W.H. (ed.) 1:183-8. Surrey, England: Unwin, 1971.

111. Schneider, W.D., and Rebohle, E. "Early Diagnosis by Means of Flow Volume Curves in Workers with Respiratory Exposure." Z. Erkr. Atmungsorgane, 157(3):291-6 (1981).

112. Seerup, K.K., and Juhl, M. "Occupational Burns. Frequency, Causes and Distribution in Occupational Groups." Ugeskr. Laeg., 143(21):1356-60 (1981).

113. Sheers, G., and Coles, R.M. "Mesothelioma Risks in a Naval Dockyard." Arch. Env. Health, 35(5):276-82 (1980).

114. Sjogren, B. "Work Environment Problems in Welding. No. 13. Relationship between Air and Urine Concentrations of Fluorides, Chromium and Nickel in Welding." Arbete och Halsa, (9):1-25 (1981).

115. Sjogron, B. "A Retrospective Cohort Study of Mortality among Stainless Steel Welders." Scand. J. Work Environ. Health, 6(3):197-200 (1980).

116. Sliney, D.H., Moss, C.E., Miller, C.G., and Stephens, J.B. "Semintransparent Curtains for Control of Optical Radiation Hazards." Applied Optics, 20(14):2352-66 (1981).

117. Smars, E. "The Problem of Ozone during MIG Welding of Aluminum." Schweiz. Alum. Rundsch., 31(6):255-61 (1981).

118. Spelbrink, H. "Preventative Health Care Examinations for Welders - Some Industrial Health Care and Administrative Aspects." Schweissen Schneiden, 34(2):84-6 (1982).

119. Spelbrink, H. "Measures in the Field of Medicine Aimed to Improve the Working Conditions of Welders." Annual Welding Conference, Sept. 16-23, 1981, Essen, West Germany. pp. 16-20. Dusseldorf, West Germany Deutscher Verband fur Schweisstechnik, 1981.

120. Stern, R.M. "Assessment of Risk of Lung Cancer for Welders." Arch. Environ. Health, 38(3) (1983) in press.

121. Stern, R.M. "Occupational Health Risk Assessment: A Preparatory Study of the Exposure of Welders to Toxic Substances and the Resulting Health Effects: The Working Environment Research Group." U.S. NTIS, PB82-119041. 555pp. (c1981).

122. Stern, R.M., "Process-Dependent Risk of Delayed Health Effects for Welders." Environ. Health Perspect., 41:235-53 (1981).

123. Stern, R.M., "A Protocol for the Study of Health Effects of Welding: The Working Environment Research Group." U.S. PB82-119421. 33pp. (c1981).

124. Sunderman, F.W., Jr. "Carcinogenic Effects of Metals." Fed. Proc., 37(1):40-6 (1978).

125. Suzuki, Y., Toda, K., Koike, S., and Yoshikawa, H."Cadmium, Copper and Zinc in the Urine of Welders Using Cadmium Containing Silver Solder." Ind. Health, 19(4):223-30 (1981).

126. Taylor, H.R. "Ultraviolet Radiation and Pterygium." J. Am. Med. Assoc., 247(12):1698 (1982).

127. Taylor, H.R., "Aetiology of Climatic Droplet Keratopathy and Pterygium (letter)." Br. J. Opthamol., 64:154-63 (1980).

128. Thomsen, E., and Stern, R.M. "Collection, Analysis and Composition of Welding Fumes." U.S. NTIS, PB82-155862. 34pp. (c1981).

129. Thomsen, E., and Stern, R.M. "A Simple Analytical Technique for the Determination of Hexavalent Chromium in Welding Fumes and Other Complex Matrices." Scand. J. Work Environ. Health, 5(4):386-403 (1979).

130. Tredre, B.E., and Collins, J.C. "A Tracking Task Used for the Simulation of Manual Metal Arc Welding." Ergonomics, 23(4):401-4 (1980).

131. Tsyrkunov, L.P. "Skin Diseases in Electric Welders." Vestn. Dermatol. Venerol., (10):40-3 (1981).

132. Tumakov, A.Z., and Grigorev, I.I. "Prevention of Meteorotropic Reactions in Workers in Various Occupations Undergoing Treatment in a Sanitarium Preventorium." Gig. Tr. Prof. Zabol., (4):22-6 (1981).

133. Ulfvarson, U. "Survey of Air Contaminants from Welding." Scand. J. Work Environ. Health, 7(Suppl. 2):1-28 (1981).

134. Vallyathan, V., Bergeron, W.N., Robichaux, P.A., and Craighead, J.E. "Pulmonary Fibrosis in an Aluminum Arc Welder." Chest, 81(3):372-4 (1982).

135. Villaume, J.E., Wasti, K., Liss-Suter, D. and Hsiao, S. *Effects of Welding on Health, Volume I.* Miami, FL: American Welding Society, 1979.

136. Weymueller, C.R. "Uncle Sam Surveys Welding Safety." Weld. Des. Fabr., 53:138-9 (1980).

137. White, L.R., Hunt, J., Richards, R.J, and Eik-Nes, K.B. "Biochemical Studies of Rat Lung Following Exposure to Potassium Dichromate or Chromium-Rich Welding Fume Particles." Toxicol. Lett., 11(1-2):159-63 (1982).

138. White, L.R., Hunt, J., Tetley, T.D., and Richards, R.J. "Biochemical and Cellular Effects of Welding Fume Particles in the Rat Lung." Ann. Occup. Hyg., 24(1):93-101 (1981).

139. White, L.R., Richards, R. J., Jakobsen, K., and Oestgaard, K. "Biological Effects of Different Types of Welding Fume Particulates." Proc. Meet., 1979. The In Vitro Effects of Mineral Dusts, Brown, R.C. (ed.) pp. 211-8 London: New York: Academic Press, 1980.

140. Wulf, B., and Seefeld, G. "Comparative Studies of the Periodontal State in Welders." Stomatol DDR, 30(7):519-26 (1980).

141. Yager, J.W. and Benz, R.D. "Exposure to Ethylene Oxide at Work Increases Sister Chromatid Exchanges in Human Peripheral Lymphocytes." Science, 219:1221-3 (1983).

142. Yatsenko, K.S., Trubnikov, G.A., Afanasev, Y.A., and Molokanov, V.M. "Vibration Sensitivity in Workers Associated with Shipbuilding." Sov. Med., (5):78-80 (1981).

143. Zakhari, S. and Anderson, R.S. *Effects of Welding on Health, Volume II.* Miami, FL: American Welding Society, 1981.

144. Zakhari, S., and Strange, J. *Effects of Welding on Health, Volume III.* Miami, FL: American Welding Society, (In press).

145. Zober, A. "Possible Dangers to the Respiratory Tract from Welding Fumes--Methods of Approach in an Industrial Health Care Context and Results." Schweissen Schneiden, 34(2):77-81 (1982).

146. Zober, A. "Symptoms and Findings at the Bronchopulmonary System of Electric Arc Welders. 1. Epidemiology." Zentralbl. Bakteriol., 173(1-2):92-119 (1981).

147. Zober, A. "Symptoms and Findings at the Bronchopulmonary System of Electric Arc Welders. 2. Pulmonary Fibroses." Zentralbl. Bakteriol., 173(1-2):120-48 (1981).

148. Zugerman, C. "Chromium in Welding Fumes." Contact Dermatitis, 8(1):69-70 (1982).

Bibliography

S & H Research Reports

WE	Welding Environment
FGW	Fumes and Gases in the Welding Environment
AWN	Arc Welding and Cutting Noise
EWH1	Effects of Welding on Health, Vol. 1
EWH2	Effects of Welding on Health, Vol. 2
EWH3	Effects of Welding on Health, Vol. 3
ULR	Ultraviolet Reflectance of Paint
WFC	Welding Fume Control
WFDP	Welding Fume Control - A Demonstration Project
LVOS	Laboratory Validation of Ozone Sampling with Spill Proof Impingers

S & H Standards

Z49.1-73	Safety in Welding and Cutting
A6. 2-73	Lens Shade Selector (wall chart 11 x 17)
F1.1-76	Method of Sampling Airborne Particulates Generated by Welding and Allied Processes
F1.2-79	Laboratory Method for Measuring Fume Generation Rates and Total Fume Emission of Welding and Allied Processes
F1.3-83	Evaluating Contaminants in the Welding Environment - A Sampling Strategy Guide
F2.1-78	Recommended Safe Practices for Electron Beam Welding and Cutting
F4.1-80	Recommended Safe Practices for the Preparation for Welding and Cutting Containers and Piping That Have Held Hazardous Substances
F6.1-78	Method for Sound Level Measurement of Manual Arc Welding and Cutting Processes
AWS	Arc Welding Safety (information booklet)